50 WAYS THE WORLD COULD END

Alok Jha

Quercus

First published in Great Britain in hardback as
The Doomsday Handbook in 2011
by Quercus Editions Ltd

This paperback edition published in 2014 by
Quercus Editions Ltd
55 Baker Street
Seventh Floor, South Block
London
W1U 8EW

A CIP catalogue record for this book is available
from the British Library

ISBN 978 1 78206 946 1

10 9 8 7 6 5 4 3 2 1

Text designed and typeset by Ellipsis Digital Ltd, Glasgow

Printed and bound in Great Britain by Clays Ltd, St Ives plc

CONTENTS

HUMAN THREATS

MASS EXTINCTION

There is a disturbing secret about the fortune of life on Earth. It sits in the background as we marvel about the diversity of organisms we see around us today; about how some chemicals in a warm pond started to replicate billions of years ago and evolved, slowly, into trilobites, dinosaurs, trees, snails, grass, monkeys, mushrooms and humans. That secret is death.

In the stunning 3.5-billion-year history of life on Earth, of the 4 billion species that are thought to have ever existed, 99 per cent are extinct. On our planet, extinction is the norm. Throughout history, plants and animals have come and gone as the climate and the environment have changed around them. Some thrived in cold, others only in hot and humid. Species have appeared and disappeared as their suitability for their niches has waxed and waned, all of it inextricably linked to the fortunes and conditions of our shared planet.

Five times in the past 500 million years, though, the steady rate of attrition has reached unassailable heights. Something, no one knows for sure what, turned the Earth into exactly the wrong planet for life, and at those points, most of the world's plants and animals disappeared. During each mass extinction, more than 75 per cent of living species died off in the geological blink of an eye.

Since the end of the last mass extinction, 65 million years ago, things have been relatively quiet. Extinctions have continued, but only at the normal, background rate. Evidence from ecologists and conservationists in recent decades, however, is pointing towards an inevitable conclusion for the 21st century: that the Earth is getting ready for another of its periodic catastrophic destructions of life. The sixth mass extinction is upon us, and the twist this time is that we know exactly what is causing it.

The five big ones

In 1982, David M. Raup of the Field Museum of Natural History in Chicago and Jack Sepkoski of the department of geophysical sciences at the University of Chicago published a study in the journal *Science*. They had been looking at the rate and deposition of thousands of families of marine fossils from the past 500 million years and noticed a distinct set of periods when extinction rates among the life forms seemed unusually high. Later work confirmed Raup and Sepkoski's conclusions that there had been some event or series of events that had caused catastrophic loss of life on Earth at several distinct points.

Because of the timescales and uncertainties in the numbers of animals involved, big extinctions in the past are not measured only by the numbers of species that disappeared, but also by genera (the taxonomic category that contains several related species) and families (the category that contains several related genera).

MASS EXTINCTION

The oldest mass extinction identified by Sepkoski and Raup started around 450 million years ago as glaciers were forming on the Earth, locking up water and causing sea levels everywhere to drop. This Ordovician-Silurian event lasted around 10 million years and caused the death of a quarter of marine families and 60 per cent of water-based genera. The hardest-hit species included brachiopods, eel-like conodonts and trilobites.

The late Devonian mass extinction came next, a 25-million-year-long event that started around 375 million years ago, and in which around 19 per cent of all families, 50 per cent of all genera and 70 per cent of all species died out. During this time, insects and plants and the first early versions of amphibians were living on land – the extinction gave them all a setback.

After that came the big one: the Permian-Triassic event. Starting around 250 million years ago, it is widely known to be the Earth's worst mass extinction event, killing 95 per cent of species living at the time, 84 per cent of marine genera and an estimated 70 per cent of land species, including plants and vertebrates. A third of the world's insects also perished, marking the only mass extinction of this order of life. The loss of life was so catastrophic that palaeontologists call this period the 'Great Dying'. On land, the mammal-like reptiles were wiped out, and it took the vertebrates tens of millions of years to bounce back. What caused this event is disputed – some argue that it was due to a comet or asteroid impact, though no tell-tale crater has ever been found.

In comparison to the Great Dying, the final two events seem less impressive, though each one had its own important influences on how life subsequently evolved on Earth.

The End-Triassic event began 214 million years ago and lasted 10 million years, possibly caused by lava flows erupting from somewhere in the mid-Atlantic region. This led to the opening of the Atlantic Ocean, and would also have caused deadly global warming, wiping out 22 per cent of marine families and 52 per cent of genera. On land, this event left the dinosaurs with little competition.

And then there is the mass extinction that most people might have a passing knowledge of, the one that wiped out the dinosaurs. Occurring around 65 million years ago, the Cretaceous-Tertiary event killed off 16 per cent of marine families, 47 per cent of marine genera and 18 per cent of land vertebrate families. The probable reason was a large asteroid impact on the Yucatan Peninsula and beneath the Gulf of Mexico, which threw dust into the air and blocked out the Sun, causing large animals to die over the next few million years due to lack of food. Afterwards, mammals and birds emerged as the dominant forms of life.

The sixth mass extinction

These days, we are used to extinction. Newspaper headlines are filled with stories about animals and plants big and small dying out or disappearing. And the rate seems to be getting ever quicker: we know that tens of thousands of species die out every year, many of them not properly catalogued by humans. We have fished the oceans empty and cleared and hunted through so much forest that our closest animal cousins, the great apes, are endangered in the wild. We are outcompeting every animal and plant on the planet by taking up an ever-increasing amount of land and energy. Our rising population (estimates suggest that we will increase from our present 6 billion or so to 9 billion by the middle of the century) means that competition will only get more intense, and it is animals rather than humans that will suffer most.

To work out how current extinction rates of species compare to those in the Earth's mass-extinction history, Anthony D. Barnosky, a biologist at the University of California, Berkeley, collected together data on species that were under threat or endangered.

Writing in *Nature* in March 2011, Barnosky described how scientists were increasingly recognizing modern extinctions of species and populations. 'Documented numbers are likely to be serious underestimates, because most species have not yet been formally described. Such observations suggest that humans are now causing the sixth mass extinction through co-opting resources, fragmenting habitats, introducing non-native species, spreading pathogens, killing species directly, and changing global climate.'

Setting a relatively high bar for what he counted as a mass extinction, in other words where more than 75 per cent of all species per group of life form were lost, Barnosky calculated that the Earth would reach its next extreme, the sixth mass extinction, in just a few centuries. 'How many more years would it take for current extinction rates to produce species losses equivalent to Big Five magnitudes? The answer is that if all "threatened" species became extinct within a century, and that rate then continued unabated, terrestrial amphibian, bird and mammal extinction would reach Big Five magnitudes in 240 to 540 years (241.7 years for amphibians, 536.6 years for birds, 334.4 years for mammals),' he wrote.

If extinction were limited to only the 'critically endangered' species over the next century, and those extinction rates continued, the time until 75 per cent of species were lost per group would be 890 years for amphibians, 2,265 years for birds and 1,519 years for mammals. At the slower end of Barnosky's calculations, the range of extinction times for amphibians is around 4,500 years, just over 11,300 years for birds and 7,500 years for mammals. 'This emphasizes that current extinction rates are higher than those that caused Big Five extinctions in geological time; they could be severe enough to carry extinction magnitudes to the Big Five benchmark in as little as three centuries.'

A separate analysis by scientists at the University of York and the University of Leeds examined the relationship between climate and biodiversity over the past 520 million years and uncovered an association between the two. When the Earth's temperatures are in a 'greenhouse' climate phase, they found that extinction rates are relatively high. Conversely, during cooler 'icehouse' conditions, biodiversity increases.

Their results, published in 2007 in the *Proceedings of the Royal Society B*, suggest that the predicted rapid rise in the Earth's temperature due to man-made climate change could halve the number of species on the planet. According to the Intergovernmental Panel on Climate Change, global temperatures could increase by as much as 6°C by the end of the century.

Could the sixth mass extinction be avoided?

Unlikely. 'Recovery of biodiversity will not occur on any timeframe meaningful to people,' says Barnosky. 'Evolution of new species typically takes at least hundreds of thousands of years, and recovery from mass extinction episodes probably occurs on timescales encompassing millions of years.'

Past mass extinctions have been caused by major environmental events – shifting sea levels, asteroid impacts or rapid temperature changes. This time, however, animals and plants have something far more formidable and insurmountable to deal with as they die in numbers never seen before on Earth: us.

GLOBAL PANDEMIC

In the winter of 2009, the world was on edge. A strain of swine flu had emerged out of nowhere in Mexico and was advancing fast. As the authorities battled to constrain it, a concern hung in the air: what if this bug was that nightmare, the global killer?

The virus spread to the US, then reached Europe within days and, within weeks of the first detection, the World Health Organization (WHO) had announced a full-blown pandemic. There is a circulating strain of influenza every year but this one, a version of H1N1, seemed more virulent than most, quicker to spread.

Four years earlier, an even nastier bug, avian influenza H5N1, had done the rounds on the other side of the world. Health authorities had been on high alert and wild birds were watched with suspicion – any of them could have harboured the bug, which had started infecting and killing humans in east Asia.

What these and other global pandemics have in common is the speed with which they can travel. Both of the scares described above tailed off within months and the spread was contained without the whole world being placed in danger.

But scientists, doctors and public health authorities are quietly worried about the big one, the nightmare virus that cannot be stopped and infects people with ease. The one that kills millions, or even billions.

In the end, swine flu H1N1 caused more than 18,000 deaths, according to the WHO, and by the end of 2005, H5N1 had killed seventy-four people. For reference, both these figures pale in comparison to the expected number of deaths from flu – between 250,000 and 500,000 every year.

What was worrying experts, in the months during which the virus was spreading, was the memory of the worst pandemic on record. In 1918, the world was recovering from four devastating years of war when the infamous Spanish influenza struck. Within six months, this virus had killed around fifty million people, with most of the dead aged between twenty and forty-five. The 1957 and 1968 Asian flu epidemics also killed millions. Today, travel is even faster and most people live in densely packed cities. A 1918-type pandemic would be devastating.

How diseases spread

You wake up with a mild headache, nothing serious enough to keep you away from work. Over breakfast you chat to your children about the day ahead and kiss them goodbye as they go off to school. On the train to work you can't find a seat and everyone is jammed together. At the office, it's a full day of meetings.

By the time the winter evening draws in, you're shivering and your nose has been running all afternoon. The next day, you recognize the signs of flu and decide to stay in bed.

But how many dozens of people have you already come into contact with? How many did you infect? And how many of them then passed the virus on to others? And that is just in a single city: how many of the people you infected went to other cities or hopped on a plane to another country? Within a day, one case could have become a dozen. In a week, thousands could be infected around the world.

It's not just death that causes problems

You might be forgiven for thinking that any modern anxiety about pandemics is hyperbole. There have been major pandemics throughout human history (the Black Death in 1348 killed a third of Europe's population), but civilization did not end. And we have better drugs today too. Why the worry?

The reason is that modern life is so much more interconnected today than it ever has been in the past. We rely on global supply chains for everything from food to pharmaceuticals. Knock out any important part of that web, and you're in trouble.

Take lorry drivers, for example. In a relatively mild pandemic, drivers might be laid up or have to look after sick family members. What if schools were closed too and they had to babysit their children? Remove a significant number of lorries from the roads and supplies do not go anywhere. In the past, shops might have kept warehouses filled with stock – everything from food to antivirals. But maintaining such a store is expensive, and improved information technology and logistics have meant that shops now rely on smaller, more frequent deliveries.

The US authorities recommend keeping three weeks' food and water stockpiled in case of a pandemic: most cities around the world have only a few days' worth of food.

What about hospitals? These need regular deliveries of drugs, blood and sterile supplies. Ambulances need fuel. And someone needs to transport fuel from refineries to petrol stations. What if none of the people required to move things around turned up for work?

What about the fuel needed for the national grid? Half of US electricity comes from coal-fired power stations, and if a pandemic stopped train drivers from moving that coal from quarries to the generators, the country's lights would not stay on for long.

When the electricity is out, we get into a whole new slew of problems. Refrigerators cannot work, so food goes off. People will not be able to cook. Radio and TV stop broadcasting. Phones and computers have to be switched off, putting critical infrastructure (including financial records) out of reach.

In 2006, economist Warwick McKibbin of the Lowry Institute for International Policy in Sydney, Australia, modelled the effects on the modern world of a 1918-style pandemic. 'The mild scenario, estimated to cost the world 1.4 million lives, reduces total output by nearly 1% or approximately $330 billion (in constant 2006 prices) during the first year,' he said. 'In our model, as the scale of the pandemic increases, so do the economic costs. A massive global economic slowdown occurs in the next-worst scenario, with more than 142 million people killed and some output in economies in the developing world shrinking by half. The loss in output in this scenario could reach $4.4 trillion, 12.6% of global GDP in the first year.'

The relatively small SARS outbreak of 2002, for example, spread across twenty-six countries in just a few months, infecting more than 8,000 people and killing more than 700 with a pneumonia-like illness. However, it had disproportionate knock-on costs, as flights were cancelled, schools shut down, and panic gripped Asian markets; it cost the world $40 billion. 'Investing in poverty reduction and healthcare in developing countries are the keys to managing pandemics in the long term,' said McKibbin. 'For now, we will have to live with a world where a relatively minor flu outbreak in Mexico City can send markets reeling in Tokyo.'

The diseases to worry about

The reason the H5N1 virus of 2005 was not as dangerous as feared, despite its virulence, was that it could not pass easily between humans. Those who died lived in close quarters with poultry. But it would not take much for the virus to mutate or combine with a human flu virus, and things would get deadly very quickly.

And flu is not the only thing that should be on our list of worries. Spotting the next big problematic virus requires monitoring of wild animals and the people who come into contact with them. 'We believe such eavesdropping may provide the early warning needed to stop pandemics before they start,' says Nathan Wolfe, a biologist at Stanford University and director of the Global Viral Forecasting Initiative (GVFI). 'Spotting new threats can happen only if scientists know what viruses are circulating among the species most likely to give rise to new pandemic viruses – birds and pigs. And whereas surveillance in the former has improved over the past five or six years thanks to concerns about bird flu (the H5N1 virus), scientists know too little about the viruses that infect the estimated 941 million domesticated pigs around the world.'

Had scientists been watching how hunters were interacting with wild animals thirty years ago, they might have been able to catch HIV early, before it reached the pandemic state, says Wolfe. The question now is, how can we prevent the next big killers? Wolfe proposes listening to what he calls viral 'chatter' – the pattern of transfer of animal viruses to humans – in the hope of sounding the alarm about an emerging infectious disease before it becomes deadly.

Scientists with the GVFI already follow people and animals in Cameroon, China, the Democratic Republic of the Congo, Laos, Madagascar and Malaysia – all hot spots for emerging human infectious diseases. More than half of such diseases, past and present, including influenza, SARS, dengue and Ebola, originated in animals. Analysis of blood from hunters and hunted has already revealed several animal viruses not previously seen in humans, including one called the simian foamy virus, which is in the same family as HIV.

None of this, by the way, takes into account synthetic viruses created by people hell-bent on terror. Genetically modified organisms bred to be virulent and also resistant to drugs might be difficult to make and disseminate today, but that will not always be the case.

What can you do?

A few months after the H1N1 pandemic died down, Peter Sandman, a risk-communication consultant from Princeton, wrote a commentary in the journal *Nature* on the lessons people could learn from the outbreak: 'For the ordinary citizen, the US government has so far recommended only hygiene. It has told people to stay at home if they are sick and to wash their hands. It hasn't told people to stock up on food, water, prescription medicines or other key supplies,' he said. 'Here is a secret of preparedness that is easy to forget: it is calming to prepare. Having things to do gives people a sense of control. It builds confidence, and it makes them more able to bear their fear.'

Because you can be sure that a virulent pandemic will happen again one day. 'Influenza was the twentieth century's weapon of mass destruction, killing more than the Nazis, more than the atomic bomb, and more than the First World War,' says John Oxford, a virologist at the Royal London School of Medicine and Dentistry. 'We would do well to dwell very seriously indeed on this fact. Nature is the greatest bioterrorist of our world, and we should concentrate and expand our efforts in public health. Emerging viruses could do for us all, as easily and as quickly, or even more so, than the Great Influenza of 1918.'

THE DOOMSDAY MACHINE

The Cold War was the closest the world ever came to a man-made Armageddon. In the decades after the Second World War, the United States and the USSR built thousands of missiles bristling with nuclear bombs, and aimed them at each other's major cities. Letting one loose would have been a disaster; firing all of them would have destroyed the world.

Both sides knew the consequences of an all-out nuclear war. And it is clear now that neither of them had the appetite to start proceedings. But at the time, they had to remain defiant and strong in the face of the enemy, seemingly ready to strike in response to any transgression. It was a time a time of bluff, counter-bluff and deep paranoia.

Any nuclear strike would, of course, have come as a shock to the other side. A radar operator might see a red dot on a screen, moving steadily closer to their country's capital. They would then have only a few minutes to make a terrifying decision. Is this a real attack? Do we strike back? What if we launch all our missiles in retaliation and the red dot turns out to be a flock of geese?

The Soviets thought hard about the problem of retaliation and came up with a system that bought them extra time in the event of seeing that red dot on their radar screens. If, for whatever reason, they could not confirm whether or not they were under attack, they would transfer their missile-launching power to an automatic system called the Perimeter, also known as Mertvaia Ruka (Dead Hand). This doomsday machine was designed to wait until an attack had been confirmed, and then, even if all the Soviet leaders had been wiped out, to initiate a devastating retaliatory strike. It would be the end of the world as we know it.

Letting the machines take over

In Stanley Kubrick's film *Dr Strangelove*, a rogue US nuclear bomber decides to attack the USSR. In the course of the story, the Americans are told by the Russian high command that if the bomber is successful, there is nothing they can do to prevent an all-out retaliation from the automatic Soviet doomsday machine – fifty hydrogen bombs positioned around the globe that will explode if they detect an attack on Russian soil. These would release enough deadly cobalt fallout to render the Earth uninhabitable for ninety-three years. And no human, not even the Russians, would be able to stop it.

No one would build such a thing for real, would they? In fact, they did. Perimeter was designed by the Soviet military to ensure the launch of all its nuclear missiles if its commanders were put out of action during a war. If the Russians were attacked by surprise and their command centres compromised, they would still be able to wreak revenge.

'The Perimeter system is very, very nice,' said Valery Yarynich, a former Soviet colonel and a thirty-year veteran of the Soviet Strategic Rocket Forces and General Staff, speaking to *Wired* magazine in 2009. 'We remove unique responsibility from high politicians and the military.'

David Hoffman, a former journalist for the *Washington Post* and the paper's Moscow bureau chief, wrote about Perimeter in a book called *Dead Hand: The Untold Story of the Cold War Arms Race and its Dangerous Legacy*. He said the machine created 'an alternative system so that the leader could just press a button that would say: I delegate this to somebody else. I don't know if there are missiles coming or not. Somebody else decide.'

In this scenario, military leaders would flip on a system that would send a signal to a deep underground bunker where three duty officers sat. 'If there were real missiles and the Kremlin were hit and the Soviet leadership was wiped out, which is what they feared, those three guys in that deep underground bunker would have to decide whether to launch very small command rockets that would take off, fly across the huge vast territory of the Soviet Union and launch all their remaining missiles.'

How would it have worked?

Perimeter monitored the ground and air around the USSR for signs of a nuclear explosion, using seismic, radiation and air-pressure sensors. If it determined that a nuclear weapon had exploded on Soviet soil, it would first check to see if the communications links to the Soviet General Staff were intact; in other words, whether anyone was still alive and in charge. If the lines were dead, it would assume that the country was under attack and that no one was left to launch a retaliatory strike. It would then re-route command of the Soviet arsenal of missiles to a secret bunker, deep underground, designed to survive an all-out nuclear attack. In that bunker, a duty officer would have complete control to launch missiles, bypassing the normal layers of command.

If the duty officer, who might be someone senior (though that was not guaranteed), decided that the USSR had been overrun by nuclear weapons, he could launch a set of control missiles that sat in armoured silos around the country. These would rise high into the sky and send down signals to the remaining Soviet nuclear missiles that the USSR had survived the wave of US nuclear warheads. They would also beam down orders to initiate the automatic launch of the Soviet warheads and rain down destruction on the US aggressors.

Perimeter was designed to lie dormant until someone high up in the military switched the system on, during a time of extreme emergency. Many former members of the Soviet military have confirmed its existence: in the 1990s, members of the Central Committee of the Communist Party gave representatives of the American defence company Braddock, Dunn & McDonald details of the Soviet preparedness for a nuclear attack.

General Varfolomei Korobushin, former deputy chief of staff of the Soviet Strategic Rocket Forces, said that the country's biggest fear in the Cold War was of a US first strike, so their main objective was to design a system that was capable of launching as soon as an attack was detected. 'Right now we have a system in place which would automatically launch all missiles remaining in our arsenal even if every nuclear command center and all of our leaders were destroyed,' he said. 'This system, called the Dead Hand, would have been triggered by a combination of light, radioactivity, and overpressure, and would cause several command rockets to be launched into orbit, from where they would send launch codes to all our remaining missiles.'

Vitalii Kataev, senior adviser to the chairman of the defence industry, explained that with the Dead Hand mechanism, the decision-maker at the centre simply unblocked the no-fire mechanism, thereby releasing launch control to local automatic triggers associated with each command missile. 'The triggers, fed by numerous sensors, will launch its local command missile and, in turn, its associated cluster of ICBMs once the sensors are excited by the light, or seismic shock, or radiation, or atmospheric density associated with an incoming nuclear strike.'

He added that the system was 'definitely operational by the early 1980s. It is important to understand that unblocking of Dead Hand assumes the scenario of a situation that is extremely threatening to the political and military leadership of the state. That basic expectation is that all decision makers are dead when the command missiles automatically fire.'

Why was it built?

It is a horrifying idea that something like Perimeter might have existed, ready to wreak further havoc even when its country had been brought to destruction. But it became operational at a time when the Soviets seemed convinced that the US would attack first.

In the 1980s, Ronald Reagan started talking publicly about his Strategic Defence Initiative (later nicknamed Star Wars), a system of lasers and explosives that would protect the US from nuclear attack. After the relative calm of the 1970s, the fact that the American president seemed unafraid of such an attack was a surprise to the Russians. The new US administration stepped up the rhetoric, playing up the country's strengths and playing down the more apocalyptic scenarios of a post-nuclear fallout.

Reagan maintained that Star Wars was purely defensive, but officials in Moscow suspected that in reality the Americans were preparing an attack. The Soviets were not keen to jump the gun and attack pre-emptively, so Perimeter became a way for them to be sure that theirs would not be the only country destroyed if the US launched their arsenal. If nothing else, it was a good deterrent.

But as any military analyst knows, a deterrent is only effective if your enemy knows what you will do in the event of an attack. Unfortunately, there is no evidence that any senior US officials knew anything about Perimeter. This made it a strategy in which even the mad scientist in Kubrick's film would have noticed gaping holes.

When Dr Strangelove, played by Peter Sellers, is told about the 50 secret nuclear bombs covered in cobalt waiting to explode automatically if the USSR is attacked, he exclaims: 'Yes, but the . . . whole point of the doomsday machine is lost if you keep it a secret! Why didn't you tell the world, eh?'

Could it be used today?

Perhaps the most alarming thing about Perimeter is that it seems to be plugged in to this day. 'We don't really know if there's still a switch in the Kremlin,' said Hoffman in an interview with Terry Gross on National Public Radio in the US. 'But that aside, I think the command rockets, the bunker, the entire

Perimeter system is still there and waiting. And I think the command system part of it is still functioning . . . I've been told that that command structure may have changed. But I do know that the men in the bunker are still there. The system is still alive. It's still a command system.'

Whether Perimeter, or any other secret system like it elsewhere in the world, could be switched on in the future depends on the diplomatic relations between the ever-growing number of nuclear powers. Nuclear weapons might have seemed the best way to keep the world away from war in the past. But disarmament has never looked like such a good idea as it does today.

MUTUALLY ASSURED DESTRUCTION

Global nuclear war is supposed to be a thing of the past, a relic of the Cold War. But human desperation for power never ends and, with the emergence of new nuclear states, the threat of annihilation is ever present.

Wars are usually about each side weighing up the risks of fighting in the hope that, if they succeed, something better lies on the other side. A sacrifice is made for the good of a greater goal.

Not so with mutually assured destruction, however, a term devised during the Cold War to describe what would happen if the two superpowers, the US and the USSR, did what they were always threatening to do and launched their thousands of nuclear warheads at each other. Both sides knew that the act would be catastrophic not just for themselves, but for the planet too. The phrase says it all: if two sides go to war, both can be assured of the complete destruction of the other.

In a nuclear strike, millions of people would die in the initial blasts and firestorms where the bombs detonated. Further afield, billions more would perish from sickness and starvation in the months and years that followed, a consequence of the Earth's sudden transformation into a barren wasteland where no plant or animal could survive.

And if you thought it was something consigned to history, think again. Scientists, armed with powerful computers and sophisticated weather models, have shown that even a regional conflict (between, say, India and Pakistan) could release enough smoke into the atmosphere to devastate global agriculture for many years and lead to widespread loss of life outside those two countries. The acronym for this type of nuclear war, MAD, is apt for something with such civilization-ending potential.

The nuclear age

The first nuclear bomb exploded into public consciousness on the morning of 6 August 1945 over the Japanese city of Hiroshima. That bomb, Little Boy, was followed a few days later by Fat Man, the device dropped on Nagasaki. Both killed tens of thousands in their first few moments and cursed countless thousands more with a lifetime of radiation sickness.

The nuclear warheads were developed in the 1940s as part of the top-secret Manhattan Project at the Los Alamos National Laboratory. Scientists working for the US military had pipped rival teams working in Germany and, some argue, brought the end of the Second World War closer.

Once the principle of this devastating bomb had been demonstrated, however, it was only a matter of time before scientists around the world would want to come up with their own versions. Fortunately, however simple the concept of a nuclear bomb is, actually building one is no easy feat.

The raw material for a nuclear bomb consists of atoms that will split when nudged. Uranium, a heavy metal that comes in two isotopes called uranium-238 and uranium-235, is ideal.

Both isotopes are radioactive and have nuclei that split apart, though only U-235 will split on command, whenever it is bombarded with neutrons. When a nucleus splits, it releases some energy and also more neutrons, which can then go on to split further nuclei. If enough atoms split in one place, the chain reaction becomes self-sustaining.

The problem faced by nuclear bomb engineers is getting hold of enough U-235: you need at least 50 kg to make a bomb. Naturally occurring uranium is mostly U-238, so it needs to be weeded out and the U-235 concentrated into the sample you are collecting. This is where it starts to get difficult: for every 25,000 tonnes of uranium ore, only 50 tonnes of metal are produced, and less than one per cent of that is uranium-235. And no standard extraction method will separate the two isotopes, because they are chemically identical. Instead, the uranium is reacted with fluorine, heated until it becomes a gas and then decanted through several thousand fine porous barriers. This separates it into two types: 'enriched', which contains mostly U-235, and 'depleted', which contains mostly U-238.

The arms race

After the end of the Second World War, the US and USSR raced to build warheads. The first Soviet bomb was detonated in 1949, much earlier than the US had expected. The 1950s saw the introduction of intercontinental ballistic missiles (ICBMs), which could carry warheads thousands of miles further than strategic bomber planes.

By the 1960s, both superpowers had demonstrated space rockets, ostensibly the start of the race to get satellites and people into orbit for exploration but with the ulterior motive of showing that they could now send warheads wherever they wished in the world.

By 1982, the US had built 11,000 nuclear warheads and the USSR had built 8,000. Both could deliver the bombs via ICBMs and from submarines, which acted as a backup in case command centres were destroyed in an initial attack. Both had extensive radar set-ups to track anything anomalous coming their way. And both had been aware for several decades that if they unleashed devastation, the other side would have time to strike back with equal force.

Nuclear winter

As demonstrated at Hiroshima and Nagasaki, nuclear bombs will kill anyone and anything within several miles of their detonation. The impact to the wider world comes later, as part of the so-called nuclear winter. In the late 1980s, scientists showed that the smoke from vast fires created during a nuclear war between the US and the USSR would cloud the entire planet and absorb so much of the incoming sunlight that the Earth's surface would become cold, dark and dry. Plants would die and food supplies would soon be exhausted. Summer would be as cold as an average winter.

More advanced projections of a nuclear winter, using modern climate models and supercomputers, confirmed the sketchy ideas of the 1980s and added detail. The effects would last a decade or more, much longer than previously thought, and the smoke from even a relatively small nuclear war would be heated and driven high into the upper atmosphere for years.

Speaking at a meeting of the American Geophysical Union in San Francisco in 2006, Richard Turco of the University of California, Los Angeles, said that detonating between 50 and 100 bombs – just 0.03 per cent of the world's current arsenal – would throw enough soot into the atmosphere to create climatic anomalies unprecedented in human history. Tens of millions of people would die, global temperatures would crash, and most of the world would be unable to grow crops for more than five years. In addition, the ozone layer, which protects the surface of the Earth from harmful ultraviolet radiation, would be depleted by 40 per cent over many inhabited areas and up to 70 per cent at the poles.

Turco examined an exchange between two countries of 100 Hiroshima-sized nuclear bombs (15 kilotonnes each), a conflict he argued was well within the ability of many emerging nuclear states. The results showed that the most densely packed countries would fare worst in the aftermath of a nuclear war. Because of the recent propensity of the population to move into high-density cities, India and Pakistan could face 12 million and 9 million immediate deaths respectively, while an attack on the UK would cause almost 3 million immediate deaths.

In the 100-warhead scenario, more than 5 million tonnes of sooty black smoke would spew from the resulting firestorms. This would float to the upper atmosphere, become heated by the Sun and end up being carried around the world. The particles would absorb sunlight, preventing it from reaching the surface, which would result in a rapid cooling of the Earth by an average of 1.25°C. 'This would be colder than the little ice age, the largest climate change in human history,' said Alan Robock, a climatologist at Rutgers University who worked with Turco on the analysis.

There would also be a reduction in precipitation around the world, by about a tenth. Blocking sunlight reduces evaporation and weakens the water cycle. In the Asian monsoon regions, rainfall would drop by up to 40 per cent.

The model also showed that the smoke would stay in the upper atmosphere far longer than anyone had previously thought. Older models had assumed that it would linger for around a year, as has been observed with the dust from volcanic eruptions. However, using improved atmospheric data, the new study showed that the climate would still be suffering a decade on from the initial conflict.

Is it likely?

During the Cold War, both superpowers were obsessed with being seen as powerful, but neither wanted to be the one to trigger the end of the world. The US and Soviet political leaders had a dedicated phone line in case of emergencies, and they maintained diplomatic relations. When there were just two sides, it was easier for one to keep the other in check.

Today, the nuclear picture is more complex. Nine nations – Russia, the US, France, China, the UK, Israel, Pakistan, India and North Korea – have more than 25,000 nuclear warheads between them. Many more nations might be on the verge of developing the weapons.

Prime ministers and presidents around the world profess to aspire to a dream world in which the number of these weapons is reduced (perhaps to zero).

In July 2009, US president Barack Obama and Russian president Dmitry Medvedev agreed to drop their deployed nuclear arsenal to between 1,500 and 1,675 by 2016. Robock calculated what might happen if this 'mother load' of weapons were detonated against urban targets in the US and Russia. Hundreds of millions of people would be killed, and a whopping 180 million tonnes of soot would be thrown into the atmosphere. Average temperatures would remain below freezing for many years in major agricultural regions.

There is little that will prevent any country determined to produce nuclear weapons from doing so. And there is no sign that countries want to be told what to do by superpower nations that already hold thousands of nuclear warheads.

Writing in *Scientific American* in 2010, Robock and Owen Brian Toon, chair of the department of atmospheric and oceanic sciences at the University of Colorado in Boulder, said that it was a 'misimpression' to think that the nightmare of a nuclear winter had gone away at the end of the nuclear arms race. 'In fact, a nuclear winter could readily be produced by the American and Russian nuclear arsenals that are slated to remain in 2012. Furthermore, the increasing number of nuclear states raises the chances of a war starting deliberately or by accident.'

They continued: 'Some extremist leaders in India advocated attacking Pakistan with nuclear weapons following recent terrorist attacks on India. Because India could rapidly overrun Pakistan with conventional forces, it would be conceivable for Pakistan to attack India with nuclear weapons if it thought that India was about to go on the offensive. Iran has threatened to destroy Israel, already a nuclear power, which in turn has vowed never to allow Iran to become a nuclear state. Each of these examples represents countries that imagine their existence to be threatened completely and with little warning. These points of conflict have the potential to erupt suddenly.'

TERRORISM

There's been an explosion in the centre of the city. A dozen people are confirmed killed and scores of buildings around the blast site are damaged. Police have cordoned off the immediate area but have also evacuated a five-mile radius. Confusion reigns.

You turn on the news to see a grave presenter relaying a message from the city authorities: stay indoors, however far you are from the centre of the blast. When your partner gets home from work an hour later, she's heard rumours that the blast contained 'something nuclear'.

Later that evening, police confirm that the city has been attacked by terrorists, and the blast was the result of a dirty bomb. Powered by conventional explosives, the device has dispersed a cloud of tiny particles of radioactive caesium into the air. This toxic dust has been settling all over the city for the past few hours, spread by strong air currents.

The police spokesman says that rooftops, roads, pavements and cars within several miles of the initial blast are now likely to be covered with caesium. Air-conditioning units will have drawn some of the cancer-causing dust into buildings. He repeats the advice of the authorities, this time with even more urgency: seal your homes and stay indoors until further notice.

In the days after the blast, you are asked to leave home while clean-up crews begin the task of making the city safe for human habitation. What are the chances, though, that you or your neighbours will come back soon, or indeed ever? The bomb might not have disrupted much of the local infrastructure, but the city itself has been wrecked, possibly for good.

Three types of attack

Terrorists will attack law-abiding populations with every means at their disposal – knives, guns or explosives – to get attention for themselves and their cause. Their intent is to strike fear and keep it running high for as long as possible. To get what they want, they need citizens to believe that an attack is imminent at any time. And they have to do it without the resources or funds normally available to governments fighting military conflicts.

In such circumstances, a dirty bomb is disproportionately deadly. These are weapons whose dangerous repercussions continue long beyond any initial blast, sowing fear and uncertainty for weeks, months, possibly years afterwards.

A typical device falls into one of three broad categories: radiological, biological or chemical.

The radiological dirty bomb (described in the scenario above) is usually made of conventional explosives, such as TNT or a mix of fertilizer and fuel oil, laced with highly radioactive materials. The explosion generates heat that vaporizes the toxic payload, which is then sprayed out over a large area. 'Weapons experts consider radiological bombs a messy but potentially effective technology that could cause tremendous psychological damage, exploiting the public's fears of invisible radiation,' wrote Michael A. Levi and Henry C. Kelly, physicists at the Federation of American Scientists in Washington DC, in an analysis of the threat of bioterrorism for *Scientific American* in 2002. 'Not weapons of mass destruction but weapons of mass disruption, these devices could wreak economic havoc by making target areas off-limits for an extended period.'

Radioactive dust, such as plutonium or americium particles, can become embedded in people's lungs, emitting harmful alpha radiation (the energetic nuclei of helium atoms) for years after the initial exposure.

'Dust from a radiological weapon would remain trapped for extended periods in cracks and crevices on the surfaces of buildings, sidewalks and streets, and some would have been swept into the interiors of buildings,' said Levi and Kelly. 'Certain materials that could be used in a radiological attack, such as cesium-137, chemically bind to glass, concrete and asphalt. More than fifteen years after the 1986 Chernobyl disaster, in which a Soviet nuclear power plant underwent a meltdown, cesium is still affixed to the sidewalks of many Scandinavian cities that were downwind of the disaster.'

Getting hold of radiological material would not be difficult today. In hospitals, radium and caesium are used as sources of radiation in cancer treatments. And weapons-grade plutonium or spent nuclear fuel has become increasingly available on the black market since the end of the Cold War led to the break-up of the Soviet Union.

A biological dirty bomb would aim to spread viruses or bacteria into a population. Explosives are unlikely to be involved here, as any blast would kill the bug the terrorist wanted to spread. Instead, it could be released in an invisible cloud of particles on to a train in rush-hour traffic, or added as contamination at some point in the food chain. The technology here is not new: in 1923, scientists affiliated to the French Naval Chemical Research Laboratory detonated pathogen bombs over animals in a field at Sevran-Livry, just outside Paris, killing many of their test subjects.

The trick for would-be biological terrorists is to weaponize their bug efficiently, making it into a form that is easily spreadable. Anthrax spores are suitable here, since they can be dried and milled into precise sizes, for example, that will lodge in people's lungs long enough to be deadly.

Chemical terrorism involves releasing into the air toxic materials that can kill fast. In 1995, the Japanese cult Aum Shinrikyo released a nerve gas called sarin on to the Tokyo subway – twelve people died and 5,000 had to be treated in hospital.

Another potent neurotoxin is ricin, which forms as a by-product when castor beans are turned into castor oil. A dose the size of a grain of salt is enough to kill an adult. Infection results in fever, nausea and abdominal pain, and victims can die of multiple organ failure within a few days of exposure.

The severity of an attack depends on the sophistication of the people undertaking it – it only takes a competent chemist to create vats of a toxic nerve agent, and there are plenty of competent chemists in the world. In the case of the Aum Shinrikyo attack on Tokyo, the terrorists seemed to have planned in a hurry – their batch of sarin was impure and their method of dispersal was to puncture a bag with the tip of an umbrella. Had they been better organized, the attack would probably have killed several thousand more people.

How hard is it to make a dirty bomb?

It depends on what the terrorist wants to use. Many natural diseases could be employed as biological weapons, for example, plague, botulism and tularaemia (a plague-like disease).

The one people worry about is anthrax, a disease caused by *Bacillus anthracis*, a bacterium that mainly affects cattle and sheep but can also infect people. The external form of the disease causes sores, but the pneumonic form can kill 90 per cent of those it infects if left untreated. Victims develop the pneumonic form by breathing in fewer than 10,000 spores of anthrax between one and five micrometres in size. These spores pass through the lining of the lungs and travel to the lymph nodes, releasing poisonous chemicals as they go. Symptoms include vomiting and fever, and if antibiotics are not available, untreated patients will die of haemorrhage, respiratory failure or toxic shock within a few days.

If terrorists could get hold of anthrax spores and grind them to the right size, all they would need to do to infect millions of people would be to drop a few hundred kilograms of the stuff on to a city from a low-flying aircraft.

Putting contaminants into the food supply chain is even easier. Lawrence Wein and Yifan Liu from Stanford University, California, calculated the effects of someone pouring botulinum toxin into a milk tanker on its way to a holding tank. 'Among bioterror attacks not involving genetic engineering, the three scenarios that arguably pose the greatest threats to humans are a smallpox attack, an airborne anthrax attack, and a release of botulinum toxin in cold drinks,' they wrote in the *Proceedings of the National Academy of Sciences* in 2005.

Botulinum is a strong poison that affects nerve function (and is used in small doses in plastic surgery to smooth wrinkles on the skin). Despite pasteurization of the infected milk, the scientists found that the would-be terrorist in their scenario would infect around 568,000 people with a fatal dose of the poison within a few days.

Is it likely?

According to the Council for Foreign Relations (CFR), a non-partisan American think tank, the Bush administration was convinced back in 2002 that al-Qaeda had radioactive materials, such as strontium-90 and caesium-137, with which they could build bombs. 'In January 2003, British officials found documents in the Afghan city of Herat indicating that al-Qaeda had successfully built a small dirty bomb. In late December 2003, homeland security officials worried that al-Qaeda would detonate a dirty bomb during New Year's Eve celebrations or college football bowl games, according to *The Washington Post*,' the CFR said on its website.

The think tank added that Iraq had already tested a one-tonne radiological bomb in 1987, but had given up on the idea because the radiation levels it generated were insufficient. 'In 1995 Chechen rebels planted, but failed to detonate a dirty bomb consisting of dynamite and cesium 137 in Moscow's Ismailovsky Park. In 2002 the United States arrested an alleged al-Qaeda operative, Jose Padilla, for plotting to build and detonate a dirty bomb in an American city. In 2003 British intelligence agents and weapons researchers found detailed diagrams and documents in Afghanistan suggesting that al-Qaeda may have succeeded in building a dirty bomb. Al-Qaeda detainees in American custody claim such a dirty bomb exists, but none have been discovered.'

Levi and Kelly also pointed to the International Atomic Energy Agency, which in 2001 stated that almost every nation in the world had the radioactive materials needed to build a dirty bomb, and that more than a hundred countries lacked the controls to prevent the theft of these materials.

If the question is whether a dirty bomb will one day go off somewhere, spreading radiological, chemical or biological material, then the answer has to be yes. It only takes a patient, skilled scientist to prepare the equipment and the active ingredient. But would such an event bring civilization close to the edge? That is less likely. A dirty bomb would no doubt have disproportionate effects, emptying cities and causing devastating psychological and economic damage. But a would-be terrorist would have to detonate hundreds, at the same time, all over the world, for his actions to have a cumulative effect. End of the world from dirty bombs? Unlikely.

DEATH BY EUPHORIA

In Aldous Huxley's *Brave New World*, people are born into specific roles and spend their lives fulfilling preordained tasks for the World State. Social mobility is just not an option in this extreme portrayal of society. Yet, the extreme loss of freedom does not lead to revolution. All thanks to a government-controlled drug called soma.

There is no war in the World State, there are no divided allegiances, and everyone knows that society comes before individuals. If things become bad in the carefully controlled world, soma is there to provide a holiday from the facts, to get things back to normal. Mustapha Mond, the Resident World Controller for Western Europe, claims that soma is there to 'calm your anger, to reconcile you to your enemies, to make you patient and long-suffering. In the past you could only accomplish these things by making a great effort and after years of hard moral training. Now, you swallow two or three half-gramme tablets, and there you are. Anybody can be virtuous now. You can carry at least half your morality about in a bottle. Christianity without tears – that's what soma is.'

Back in the real world, soma does not exist. At least, not yet. We have an ever-increasing repertoire of drugs to treat the unwell, but soon there will also be drugs for the healthy, to improve mental skills and alertness, to de-stress after hellish days at work and to induce euphoria at the weekends, all without side effects. How long before we use so many drugs so willingly that we are no longer in control? Perhaps the end of society as we know it will not come with a bang, but will fade away in a self-administered medical haze.

Medicines now, medicines to come

Psychoactive drugs have been used by human societies since prehistoric times. Archaeological evidence beneath houses in north-west Peru, dating back more than 10,000 years, shows that residents used to chew coca leaves. The alkaloids in the leaves are known to be stimulating, and can mitigate the effects of high-altitude, low-oxygen living.

In the past century, and in particular the last few decades, our improved understanding of human physiology has brought an explosion of drugs (medicinal and illicit) that can help lift mood, improve alertness or just keep you awake for days on end.

Most of us take drugs every day – how many people do you know who cannot start the day without the mind-sharpening effects of caffeine or nicotine?

But healthy people are also taking plenty of prescription drugs. Methylphenidate (trade name Ritalin) is given to children with attention deficit hyperactivity disorder, but it is also used by healthy people to enhance their mental performance. Modafinil, a drug developed to treat narcolepsy, has been shown to reduce impulsiveness and help people focus on problems, because it can improve working memory as well as planning.

It has already been used by the US military to keep soldiers awake and alert, and some scientists are considering its usefulness in helping shift workers deal with erratic working hours. Propranolol, a beta-blocker, is used to treat high blood pressure, angina and abnormal heart rhythms – it is also used sometimes by snooker players to calm their nerves before a game.

Almost 7 per cent of students in US universities have used prescription stimulants for non-medical purposes, and on some campuses, up to a quarter of students have used them. 'These students are early adopters of a trend that is likely to grow, and indications suggest that they're not alone,' says Henry Greely, a professor at Stanford Law School.

In a 2008 commentary for *Nature*, co-authored with a slew of experts in ethics, neuroscience, psychology and medicine, Greely explains that a modest degree of memory enhancement has been found with ADHD drugs such as methylphenidate, as well as the drug donepezil, developed for the treatment of Alzheimer's disease. 'It is too early to know whether any of these new drugs will be proven safe and effective, but if one is, it will surely be sought by healthy middle-aged and elderly people contending with normal age-related memory decline, as well as by people of all ages preparing for academic or licensure examinations.'

A 2005 study by the UK government's science think tank Foresight looked at the future for mind-enhancing drugs. In the resulting report, leading scientists in the fields of psychology and neuroscience argued that there really would be a pill for every ill very soon. And that all of it would be possible without addiction. In a world that is increasingly non-stop and competitive, an individual's use of drugs may move from the fringe to the norm.

Trevor Robbins, an experimental psychologist at Cambridge University and one of the authors of the Foresight report, pointed to a future where drugs would be used to vaccinate against substances such as nicotine, alcohol and cocaine. They could cause the immune system to produce antibodies against the drug being abused – these would render the drug impotent when taken and prevent it from having any effect on the brain.

Other drugs might delete painful memories, which could be useful for people suffering from post-traumatic stress disorder. 'We are now looking 20–25 years ahead,' said Robbins at the launch of the report. 'Very basic science is showing that it is possible to call up a memory, knock it on the head and produce selective amnesia.'

And all of this could be done so that you get the benefits of a drug with none of the pain. The harmful side effects of today's drugs could in future simply be engineered out.

Unexpected consequences

It is hard not to get excited at the possibilities of pharmaceuticals that could solve all our problems, make life better, get rid of painful memories or just put a happy (but very safe) glow on our everyday existence.

The inevitable difficulty arises from how little we really understand of what any drug actually does to our bodies and minds. Modafinil, for example, seems to be involved in modulating several chemical messengers in the brain, but no one is really certain how it works. With an expanded use of such stimulants comes the concern that in the long run, the drugs might take a toll on the brain.

Several studies in animals have shown that stimulants could alter the structure and function of the brain in ways that could depress mood, boost anxiety and, contrary to their short-term effects, lead to cognitive deficits. In 2006, scientists at the US Food and Drug Administration collated information from studies looking at children and teenagers who took anti-depressants for depression, anxiety disorders and ADHD – they found that those children had twice the risk (4 per cent versus 2 per cent) of contemplating or attempting suicide compared with those on placebos.

The unknowns are not limited to carefully designed phar-maceuticals. Research by Robin Murray, a professor of psychiatry at the Maudsley hospital in south London and one of Britain's leading experts on mental health, shows that cannabis almost always exacerbates the symptoms of psychosis in people who are already suffering from (or have a family history of) mental health problems. A study published in the *British Medical Journal* by his colleague, Louise Arsenault, showed that people taking cannabis at the age of 18 were around 60 per cent more likely to develop psychosis in later life. 'But if you started by the time you were 15, then the risk was much greater, around 450%,' says Murray.

This does not mean that everyone who smokes cannabis will develop psychosis, but it does seem to exacerbate the problem in those who have a heavy predisposition.

Would people do it?

The takeover of our entire society by drugs would require us all to willingly take them in large quantities, on a regular basis. This is not as outlandish an idea as you might think.

For one thing, we are all living longer, thanks to the benefits of decades of medical advances. But no one knows what living to 150 or 200 might do to our minds. Perhaps it will just bring us even more years in which to feel lonely or to develop depression, something already hinted at in figures from the World Health Organization and others, which show big rises in psychological disorders in ageing populations. If we can live to 200 or more, who is to say that so many years of accumulated memories and feelings might not overwhelm our primate brains, which only evolved to survive for several decades at most? One solution might be to develop even more drugs to counteract and soothe the effects of age-related depression or decline – people would have to take these to keep themselves sane.

A less extreme scenario involves peer pressure. You may object to using mind-altering drugs to help you work harder or longer. But in a world where all your colleagues (and competitors) are doing it, can you afford to be left behind? If you are a 70-year-old office worker halfway through your life, and you have to compete for work with a well-educated 23-year-old in a developing country, would you not take every advantage you could lay your hands on?

Perhaps you want to keep your life as natural and additive-free as possible. Too late, argues Greely. 'The lives of almost all living humans are deeply unnatural; our homes, our clothes and our food – to say nothing of the medical care we enjoy – bear little relation to our species' "natural" state. Given the many cognitive-enhancing tools we accept already, from writing to laptop computers, why draw the line here and say, thus far but no further?'

Human ingenuity, he goes on to say, has given us means of enhancing our brains through inventions such as written language, printing and the Internet. Drugs should be viewed in the same general category as education, good health habits and information technology: simply as ways in which our uniquely innovative species tries to improve itself.

'Safe and effective cognitive enhancers will benefit both the individual and society,' concludes Greely. 'But it would also be foolish to ignore problems that such use of drugs could create or exacerbate. With this, as with other technologies, we need to think and work hard to maximize its benefits and minimize its harms.'

It might not be called soma, it might not even be a single drug, but our future is going to be pharmaceutical.

OVERPOPULATION

At the end of the 18th century, a young cleric named Thomas Malthus noticed that he was carrying out far more christenings than funerals in his local village church in Surrey, southeast England. That insight led him to write a dire warning against the perils of unchecked human reproduction.

Malthus believed the masses were on a treadmill of sex and procreation. In his 1798 *Essay on the Principle of Population*, he described how the poorest reproduced with such vigour that their numbers would soon be culled by disease and hunger, as the land they lived on became unable to sustain them. While human populations grew exponentially with every generation (a city of 1 million people would become 2 million in a single generation, then 4 million, then 8 million and so on), the ability to feed them could only rise arithmetically in the same time (from being able to feed 1 million to 2 million in a single generation, then 3 million the generation after that, and then 4 million). Eventually, the world would run out of food, thought Malthus. People would die of starvation. It was a terrible realization, but it was nature's way of keeping populations in check.

Fortunately for the world's people, these dire predictions turned out to be overly pessimistic. Malthus wrote his essay at the end of a millennium when European death rates had been largely determined by the success or failure of harvests, but the Industrial Revolution soon changed that. Britain no longer grew all of its own food, relying instead on its colonies – sugar came in from the Caribbean, wheat from India, tea from Ceylon and meat from Australia.

And human ingenuity did some incredible things with agriculture in the centuries after Malthus, increasing the productivity of an acre of land to way beyond anything the cleric might have thought possible. In the late 1960s, Norman Borlaug won a Nobel prize for developing high-yielding varieties of dwarf wheat that, if fed with water and fertilizer, would grow large heads without falling over. By the mid-1970s, wheat and maize yields in places such as India had doubled. Similar research around the same time pushed out large-grained 'miracle rice' in the Philippines. More people, in the end, did not necessarily mean more starvation.

But today, in the 21st century, the underlying sentiment of Malthusian angst, that at some point we will run out of space or resources, is back. The question might no longer be whether we can technologically support an ever-growing population; it is morphing into whether we should. Can the Earth bear billions more humans without falling apart?

The population conundrum

There was a time when optimists had dismissed the worst of the Malthusian nightmare scenario and many of them firmly believed that more people was a good thing. More people meant more ideas, more talent and more innovation for our world.

But today, things have changed. With pressure on climate and issues around overconsumption, especially in the West, rising population is once again becoming a metaphor for the ability of humankind to treat the world with reckless abandon. More people means less water for farmers in Africa, less land for everyone and less capacity in the atmosphere for our greenhouse gases. Not to mention fewer jobs and harder living.

'There are 6.8 billion of us today, and more are on the way,' wrote Robert Engelman, vice president for programmes at the Worldwatch Institute, in 2009. 'To make a dent in these problems in the short term without throwing anyone overboard, we will need to radically reduce individuals' footprint on the environment through improvements in technology and possibly wrenching changes in lifestyle. But until the world's population stops growing, there will be no end to the need to squeeze individuals' consumption of fossil fuels and other natural resources. A close look at this problem is sobering: short of catastrophic leaps in the death rate or unwanted crashes in fertility, the world's population is all but certain to grow by at least 1 billion to 2 billion people.'

Those countries that consume at lower levels – mostly the developing countries, but also India and China on a per capita basis – would no doubt love to match the lifestyle of the average American or Briton. But to do so would push the world over the edge.

Joel E. Cohen, in the laboratory of populations at Rocke- feller University in New York, has tracked the growth of human population through history. Over the past 2,000 years, he says, the annual rate of increase of global population grew about fifty-fold, from an average of 0.04 per cent per year between AD 1 and 1650 to its all-time peak of 2.1 per cent per year around 1965 to 1970. 'Human influence on the planet has increased faster than the human population,' he wrote in a paper in the journal *Science*. 'For example, while the human population more than quadrupled from 1860 to 1991, human use of inanimate energy increased from 10^9 (1 billion) megawatt-hours/year to 93 billion megawatt-hours/year.'

Cohen predicted that the world population would double within forty-three years (by 2038) if it continued its 1.6 per cent growth rate, though he said that was unlikely. 'The population of less developed regions is growing at 1.9% per year, while that of more developed regions grows at 0.3 to 0.4% per year. The future of the human population, like the futures of its economies, environments, and cultures, is highly unpredictable. The United Nations regularly publishes projections that range from high to low. A high projection published in 1992 assumed that the worldwide average number of children born to a woman during her lifetime at current birthrates (the total fertility rate, or TFR) would fall to 2.5 children per woman in the 21st century; in this scenario, the population would grow to 12.5 billion by 2050.'

In 1960, Heinz von Foerster of the University of Illinois took the predictions of population to a tongue-in-cheek extreme by developing a model of growth that became known as 'the doomsday equation'. He wrote in *Science* that on Friday 13 November 2026, 'human population will approach infinity if it grows as it has grown in the last two millennia'.

He based his calculations on 'conditions which come close to being paradise – that is, no environmental hazards, unlimited food supply, and no detrimental interaction between elements – the fate of a biological population as a whole is completely determined at all times by reference to the two fundamental properties of an individual element: its fertility and its mortality'.

Von Foerster's intention was to add fuel to the heated controversy about whether the time had come for something to be done about population growth control. 'This controversy has divided those elements of the population under consideration who profess to show some interest in human affairs into two strictly opposed camps: the optimists, who see in the population explosion a welcome expansion of their clientele, be it consumers of baby goods, voters, or devoted souls, and, on the other hand, the pessimists, who worry about the rapid depletion of the natural resources and the irreversible poisoning of our biosphere.'

According to von Foerster, the optimists adhere to the belief that, no matter how fast the numbers grow, food technology and industry will easily keep pace – the principle of 'adequate technology', which has proved to be correct for over 100 generations, he says, will hold for at least three more.

The pessimists, meanwhile, anticipate that further rapid increase in population density will be accompanied by 'a deterioration in human dignity, and they see the ultimate fate of the human race as a mere vegetation of the individual on the edge of existence, if no measures are introduced to keep the world population under control'.

If we can grow in numbers, should we?

Heinz von Foerster's pessimists worry about environmental degradation for a good reason. The growth of our species from humble beginnings in Africa has already put enough carbon dioxide into the atmosphere to bring us to the edge of an environmental catastrophe. In the most optimistic scenario, the world will warm by an average of at least 2°C by the end of this century, and already we can see that this is bringing harsher droughts, more intense storms and higher sea levels.

If we continue to grow in numbers to 9 billion or more by the middle of this century, that would be bad enough. But the growth is likely to come in countries that want to individually increase their per capita consumption of energy and resources. 'The same one-two punch of population growth followed by consumption growth is now occurring in China (1.34 billion people) and India (1.2 billion),' said Engelman. 'Per capita commercial energy use has been growing so rapidly in both countries (or at least it was through 2007 on the eve of the economic meltdown) that if the trends continue unabated the typical Chinese will outconsume the typical American before 2040, with Indians surpassing Americans by 2080. Population and consumption thus feed on each other's growth to expand humans' environmental footprint exponentially over time.'

Which way will it go?

Predictions are one thing, but how far population actually grows or falls in the next century will depend on a matrix of decisions based on economics, environment, culture, politics and demography. Joel E. Cohen tried to survey what the actual carrying capacity of the Earth might be, but the range of predictions that had been made by various academics and organizations turned out to be bewildering. 'Many authors gave both a low estimate and a high estimate,' he wrote. 'Considering only the highest number given when an author stated a range, and including all single or point estimates, the median of 65 upper bounds on human population was 12 billion. If the lowest number given is used when an author stated a range of estimates, and all point estimates are included otherwise, the median of 65 estimated bounds on human population was 7.7 billion.'

This range of low to high medians, 7.7 billion to 12 billion, is very close to the range of low and high UN projections for 2050: 7.8 billion to 12.5 billion. 'A historical survey of estimated limits is no proof that limits lie in this range. It is merely a warning that the human population is entering a zone where limits on the human carrying capacity of Earth have been anticipated and may be encountered,' said Cohen.

What can we do?

Is some form of population control the only way to head off Heinz von Foerster's doomsday prediction? In the UK, an all-party parliamentary group on population has called for better efforts to curb population, and arch-environmentalist Jonathon Porritt, former chair of the UK government's Sustainable Development Commission, has suggested that parents of more than two children were being irresponsible.

'Two big questions present themselves as population re-emerges from the shadows: Can any feasible downshift in population growth actually put the environment on a more sustainable path? And if so, are there measures that the public and policy makers would support that could actually bring about such a change?' says Robert Engelman.

Slower population growth would not hurt. Brian O'Neill of the National Center for Atmospheric Research in the US has calculated that if we could slow the process enough to keep the 2050 population to 8 billion rather than the projected 9.1 billion, we would save up to 2 billion tonnes of carbon annually. Plus, the billion-plus fewer people would need less land, water, fish, food and forest products.

'Nature, of course, couldn't care less how many of us there are,' says O'Neill. 'What matters to the environment are the sums of human pulls and pushes, the extractions of resources and the injections of wastes. When these exceed key tipping points, nature and its systems can change quickly and drama-tically. But the magnitudes of environmental impacts stem not just from our numbers but also from behaviors we learn from our parents and cultures. Broadly speaking, if population is the number of us, then consumption is the way each of us behaves. In this unequal world, the behavior of a dozen people in one place sometimes has more environmental impact than does that of a few hundred somewhere else.'

In his 2009 book *Peoplequake*, environment journalist Fred Pearce claimed that the problem for the environment is not an increasing number of people, but increasing consumption. He pointed out that the poorest 3 billion, around 45 per cent of the total, are currently responsible for 7 per cent of carbon dioxide emissions, while the richest half a billion, around 7 per cent, are responsible for 50 per cent of emissions. 'A rural woman in Ethiopia can have ten children and her family will still do less damage, and consume fewer resources, than the family of the average soccer mom in Minnesota or Manchester or Munich. In the unlikely event that her ten children live to adulthood, and all have ten children of their own, the entire clan of more than 100 will still be emitting only about as much carbon dioxide each year as you or me.'

Population doomsday, it seems, can be avoided. But only if we want it enough.

POPULATION DEATH SPIRAL

What happens when your country cannot replace its citizens who are leaving or dying? Fewer people means a smaller workforce and less tax income, and that means less money for basic services. It could spark the end of a country.

Too many people in the world, isn't that what we are normally told to worry about? When it comes to the implications of working out the optimum number of people on the planet, through the ages the problem has usually been one of overconsumption and declining resources.

But shrinking populations are a major headache for economic growth. As well as less tax, fewer people means less innovation and industry in a country, and it can mean reduced economic and political power on the global stage. No wonder that the countries with ageing and declining populations – big economies including Japan, Russia and Australia – are so desperate to balance things out again.

This is not a classic doomsday scenario – it is unlikely that the human race would disappear completely by not having enough babies. Multiplying, after all, is something we are quite good at.

Declining populations might not end the world in the

cataclysmic sense, but they will radically shift the balance of power. Whether it will be for good or ill is an open, and worrying, question.

What is happening to population?

The number of people in the world is going up, overall. But the rate of growth has been slowing in recent decades. Global population replacement level, the number of births required to keep population stable, is just over 2.3 babies per couple. Having said that, however, actual birth rates have been dropping around the world, thanks mainly to increased access to contraception and improving education for women.

In the 1950s, birth rates were between five and six per couple; this number fell by the late 1970s to 3.9. By 2000 it had come down to 2.8, and by 2008, it was 2.6. In his book *Peoplequake*, environment writer Fred Pearce says that more than sixty countries, containing almost half the world's population, now have fertility rates below their national replacement levels. At the current rate of decline, the world's fertility rate will be below replacement level soon after 2020.

In its 2008 projections, the United Nations Population Division pointed out that in the developed countries, the population aged sixty or over is increasing at the fastest pace ever (1.9 per cent every year) and is expected to grow by more than 50 per cent over the next four decades, rising from 264 million in 2009 to 416 million in 2050. At the same time, 'total fertility is expected to fall from 2.56 children per woman in 2005–10 to 2.02 in 2045–50 according to the medium variant. In 2005–10, twenty-five developed countries, including Japan and most of the countries in Southern and Eastern

Europe, still had fertility levels below 1.5 children per woman.'

Based on these projections, the UN says that many countries around the world would face population declines in the coming decades, including Japan, Russia, Belarus, Moldova, Estonia, Canada and Italy. Those approaching decline include Greece, Spain, Cuba and Lesotho.

Factors in the decline

Populations can fall for a variety of reasons, including emigration, war, disease, famine or forced control. The Black Death in Europe and the arrival of the Spanish conquistadors to the Americas crashed populations in those places – in the latter case as much due to a lack of immunity to European pathogens on the part of the native Americans as to the wars and slaughter that accompanied the invasions.

Worries about overpopulation have also led to population control in many parts of the world in the past century. It started with the ideas of an 18th-century English cleric, Thomas Malthus, who was worried about famine caused by unchecked population growth (see previous chapter). Ever since then, people in power have tried to keep numbers down, not always with the noblest of motivations. Malthus opposed helping the poor – he argued against vaccinations, for example, because they would boost populations. The ideas in his *Essay on the Principle of Population* went on to inspire the first eugenicists. Disowned today, but hugely influential in the first half of the 20th century, eugenics built upon dubious ideas of racial and class superiorities – if the world could only support a limited number of people, reasoned the eugenicists, better that they were educated, middle class and white.

The 20th-century version of eugenics was controlled by governments. By the 1950s, 'population controllers' were everywhere, wringing their hands in NGOs and United Nations agencies, worrying about the coming Malthusian population catastrophe, looking to the poorest parts of the world to curb population growth. Mass US-funded family-planning programmes were targeted at countries such as Turkey, Malaysia, Egypt, Chile, Morocco, Kenya and Jamaica (with foreign aid and even trade sometimes dependent on an adherence to Western demands to reduce numbers). In India, the government paid its citizens to be sterilized (sometimes coercively), while China famously enacted a one-child policy that led to brutal forced abortions for those who broke the rules.

The declines to come

Though Malthus was proved wrong in his predictions (he did not foresee the improvements in agricultural technology that would feed millions more people), environmentalists still predict population crashes in future, as a result of the new pressures of climate change. Increasingly, water-stressed areas (or those that will be uninhabitable due to flooding) will compromise crop harvests, they say, and force people to migrate for basic resources. Countries will go to war for food and water, and all the while, many millions (perhaps billions) will perish from hunger or diseases that have increased their range because of the warmer temperatures.

In the past two decades, scientists have also noted pointers towards a drop in natural fertility. Sperm counts in men, for example, are on the decline around the world by between 25 and 50 per cent, according to various studies. 'It has been presumed that they reflect adverse effects of environmental or lifestyle factors on the male rather than, for example, genetic changes in susceptibility,' says Shiva Dindyal, of the Imperial College School of Medicine. 'If the decrease in sperm counts were to continue at the rate that it is then in a few years we will witness widespread male infertility. To date it remains unknown why this is happening and the available preventative measures, which can be taken to avoid a continuation of this trend, are not common knowledge.'

One hypothesis is that sperm counts are falling due to the increasing cocktail of chemical pollutants present in our modern environment, particularly those that can mimic the female hormone oestrogen. 'These chemicals are present in the plastic lining of food cans, in pesticides, in plastics and in paints. In laboratories many designed chemicals have been shown to have oestrogenic effects,' says Dindyal. 'Oestrogenic hormones exert their many effects by binding to intracellular oestrogen receptors, which consist principally of specialised proteins located within the target cells; they recognize the hormone and allow it to regulate specific oestrogen responsive genes within the cell.' These 'false' oestrogens might stick to human cells, stopping a person's natural hormones from working properly.

Additional pressures on sperm counts could include smoking, drinking and unnecessary drug-taking. Scientists at the University of Idaho found that toxic chemicals can damage sperm, which then pass altered genes on to babies. In experiments on rats, they found that some garden chemicals lead to conditions such as damaged and overgrown prostates, infertility and kidney problems, all of which are present up to four generations later.

What can we do?

The Russians are already on to the problem. In 2006, then-president Vladimir Putin brought out a plan to offer financial incentives to women to have children. The population in Russia had been falling since the end of the Soviet Union, and was being further reduced by emigration and disease, including HIV infections. These pressures could, thought the government, reduce the overall population by a third by 2050.

Similar schemes have operated in other countries: Australian couples were offered $4,000 for every baby and got their childcare costs paid for too; France, Italy and Poland have all had bonus schemes for families.

In Japan, couples with babies in the town of Yamatsuri, just outside Tokyo, receive a bonus and yearly payments for every child for its first ten years. Singapore tops the list, though, with $3,000 for the first child, $9,000 for the second and double that for subsequent children.

Declining populations are not going to end human civilization, though they will have serious and profound implications for how societies are organized and what importance they have on the global stage. It might just be part of the natural cycle of things, as different countries take the lead depending on their demographic advantages.

Will the world change with fewer people in it? Definitely. Will it end in calamity? Perhaps only for a few.

TECH

CYBERWAR

In 2002, a group of concerned scientists wrote a letter to President George W. Bush. Their message was blunt: the US was at grave risk of suffering an attack that could damage the national psyche and the economy much more broadly than even the heinous crimes committed by terrorists on September 11 the previous year.

The entire critical infrastructure of the country, including electrical power, finance, telecommunications, health care, transportation, water, defence and the internet, was vulnerable. The scientists called for 'fast and resolute' mitigating action to avoid a national disaster. 'We, as concerned scientists and leaders, seek your help and offer ours,' they wrote.

The letter-writers came from a wide range of institutions – technology companies, the halls of academia and policy think tanks. In their message to the President, they outlined the consequences of inaction.

'Consider the following scenario. A terrorist organization announces one morning that they will shut down the Pacific Northwest electrical power grid for six hours starting at 4:00 p.m.; they then do so. The same group then announces that they will disable the primary telecommunication trunk circuits between the US East and West Coasts for a half day; they then do so, despite our best efforts to defend against them. Then they threaten to bring down the air traffic control system supporting New York City, grounding all traffic and diverting inbound traffic; they then do so.'

Other threats follow, said the scientists, demonstrating the adversary's capability to attack critical infrastructure. Finally, the terrorists promise to cripple e-commerce and credit-card servicing unless their long list of demands is met. 'Imagine the ensuing public panic and chaos. If this scenario were to unfold, Americans everywhere would feel that our national sovereignty had been compromised; we would wonder how, as a nation, we could have let this happen.'

The example was America, but they could be describing any modern country in the world. Such a disaster scenario is the consequence of a society dependent on computers, and the warning was against a new type of battle that could bring the world's economies to the edge of oblivion: cyberwarfare.

Our world, the network

Ever since the 1980s, Hollywood films have featured lone hackers taking over distant military computer networks from their bedrooms. Real-life concerns about potential damage by cyberterrorists, though, had to wait for the widespread use of the World Wide Web in the late 1980s and early 1990s. Before then, private computer networks (mainly military or corporate) had existed in relative isolation and were limited to sharing messages between each other. The critical infrastructure of cities lay in physical human hands; any computers present were there to help with information rather than control.

We all know what has happened since. The price of computer chips has fallen every year, computers become ever more powerful and we have given over control of almost everything to the machines. In addition, every one of them is linked into the Internet. Computers keep our modern world turning, relaying information and commands to each other in the background, monitoring power stations, operating traffic lights, keeping planes safely in the air and chemical plants running at top efficiency.

This network has brought untold advantages to our society. But, like anything that is valuable, it has also become a prime target for anyone wanting to hold the world to ransom.

Computers have transformed modern warfare, allowing pilots in a control room in the US to operate drone aircraft thousands of miles away. Bombs are guided by GPS satellites, and fighter planes and warships have become moving data-processing centres. Each additional use of networked computers has brought advantages, but each one is a new point of attack or control.

On the civilian side, shutting down a country's critical infrastructure for a week or more would cripple it, costing the economy billions of dollars per day and leading to widespread panic and fear. If all a country's power stations were switched off from afar, or damaged in some way, how long before cities were overrun with fear and crime? Without electricity, banks cannot give out money, hospitals cannot look after the sick and building security is gone.

If that happened to lots of countries at once, we would have been hacked back into the first half of the 20th century. Which might not sound too bad until you bear in mind how much of your modern life depends on being connected though the Internet. Sending messages, contacting colleagues, looking up information and accessing your bank account is all done online. The logistics that mean that your nearest supermarket is stocked with the freshest food depend on networked computers; the balanced loads of electricity that flow to your home and office depend on networked computers. We modern humans could live without them, but we've chosen to forget how to do so.

In an article on the dangers of cyberterrorism, *The Economist* described cyberspace as the fifth domain of warfare after land, sea, air and space. To highlight the importance of the issue, President Barack Obama recently declared his country's digital infrastructure a 'strategic national asset', and the Pentagon has appointed General Keith Alexander, director of the National Security Agency (NSA), to head up a new Cyber Command to defend American military networks and attack those of other countries.

In the UK, GCHQ, the equivalent of the NSA, has a cyber-security unit, and countries including China, Russia, Israel and North Korea are known to be preparing for cyberwar.

The electronic war

Before wars became virtual, it was relatively straightforward to work out the strength of your opponent. Your enemy's main assets were missiles, tanks and soldiers. It would be easy to work out what kind of threat they posed and how quickly they could mobilize against you.

In a cyberwar, the enemy is hidden. A hacker can live anywhere in the world, hijack a computer in a second country and use that to launch attacks on a third nation. And that is the simplest scenario – teams of hackers all over the world using computers all over the world to prey on a single country's networks from a multitude of angles at the same time.

Richard Clarke, a cybersecurity and counterterrorism adviser to several successive US presidents, believes that an electronic attack could bring about a catastrophic breakdown in less than fifteen minutes. 'Computer bugs bring down military e-mail systems; oil refineries and pipelines explode; air-traffic-control systems collapse; freight and metro trains derail; financial data are scrambled; the electrical grid goes down in the eastern United States; orbiting satellites spin out of control. Society soon breaks down as food becomes scarce and money runs out. Worst of all, the identity of the attacker may remain a mystery.'

Between 2007 and 2008, computer hackers broke into the computer systems of the Joint Strike Fighter, an advanced US fighter jet, leading to the theft of details of the plane's design and electronic systems. The attack seems to have come from China, but that could have been a smokescreen by the attackers, who made off with several terabytes of information.

Estonia, a country with one of the world's most sophisticated computer networks, was the target of a series of cyberattacks in 2007. Banks, newspapers and government departments all saw their websites swamped by requests and spam, crippling the country's web access and computer networks. The attacks came from computers in more than 100 countries, but the perpetrators have never been identified.

These are just two of the attacks that we know about. Governments around the world might be planting sophisticated sleeper viruses into the networks of other countries as you read this, ready to spring into action at their master's bidding, sabotaging an enemy country's computers. And who is to say that a well-funded terror group is not doing the very same thing? Of the almost 150 billion emails sent every day, 90 per cent are known to be spam. Within those fragments of electronic communication could lurk unfathomable dangers, and it only takes one person to click a dubious email for the infection to spread through a country's network.

Computer viruses used to be all about the programmers' delight in infecting as many computers as possible: the ILOVEYOU virus in 2000, for example, caused almost $10 billion of damage by overwriting files, and infected around a tenth of the world's Internet-connected computers. Nowadays the viruses sit on computers and try to find out sensitive data such as bank details or passwords. According to *The Economist*, more than $1 trillion was lost in 2009 to cybercrime – an amount larger than the value of the world's drugs trade.

Can we do anything about it?

Like any weapon that could have huge deleterious impacts, the only check against cyberwar is human behaviour. Keith Alexander welcomes the idea of a treaty between the major economies to agree common standards and rules when engaging in cyberwarfare. 'That said, a START-style treaty may prove impossible to negotiate,' says *The Economist*. 'Nuclear warheads can be counted and missiles tracked. Cyberweapons are more like biological agents; they can be made just about anywhere.'

In their letter to President George W. Bush, the concerned scientists proposed that the problem was so important it deserved a focused plan akin to the Manhattan Project during the Second World War, which employed hundreds of scientists and engineers to build an atomic bomb to create a cyber-defence policy. 'To prevent attacks, we need a coordinated effort to work with our critical-infrastructure providers in defending their most critical information systems,' they wrote. 'To detect attacks, we need to permeate our critical networks with a broad sensor grid imbued with the capability to detect large-scale attacks by correlating and fusing seemingly unrelated events that are, in fact, part of a coordinated attack. To respond to attacks, we need to devise strategies and tactics to pre-plan effective actions in the face of major cyber-attack scenarios; we need to augment our national infrastructure with mechanisms that support the defined strategies and tactics when attacks are detected and verified.'

Some of that has been done in the decade since those scientists wrote their letter. But technology has changed too – ten years is a vast amount of time in the computer world, and the networks in place today are even more complex and woven into our lives. Citizens are connected almost all the time via their mobile phones, and data flies through the air like never before. Still, the sentiment of that letter in 2002 is as important as ever. The cybercatastrophe could come sooner than anyone thinks, and it would pay to get ready. 'The clock,' wrote the scientists, 'is ticking.'

BIOTECH DISASTER

In the past half century, our ability to manipulate plants has reached the level of messing with the most basic molecular machinery. The technology holds the potential for producing better foods but, as with anything that might become part of our environment, also the danger of unintended consequences. Understanding proteins and DNA has the potential to help scientists and farmers grow better crops: wheat that is resistant to herbicides or which can thrive in the most arid conditions; or tomatoes that stay fresh for longer and potatoes or rice supercharged with vitamins.

But when we mess with the basic elements of life, we also need to be careful. What if the transplanted genes, never meant to be in that plant in the first place, leak out into the sur-rounding, wild environment? Might the genes that confer herbicide protection to the wheat end up in the weeds on the side of the road? What if the pesticides used on those plants force the evolution of resistant bugs that spread across a country, unstoppable by even the most potent chemicals?

If that concerns you, then there is something worse in store for the coming decades too. In future, with the emerging field of synthetic biology, we will be able to build life forms from scratch, programmed to do useful things for us. But what if these techniques (creating genetically altered viruses or microbes, say) are used by terrorist groups to spread disease? If they got the formulation right, perhaps a trained evildoer could create an airborne Ebola strain that could infect the world in days.

Unlike other doomsday weapons, such as nuclear bombs, biological technology is easily available, and anyone with a makeshift research lab could do global harm. Martin Rees, former president of the Royal Society, has remarked that a million lives might be lost because of a bioterror or 'bioerror' event.

'It's scary as hell,' said Drew Endy of Stanford University, a pioneer in synthetic biology, during an interview with the *New Yorker* magazine on the potential harm of misapplied technology. 'It's the coolest platform science has ever produced, but the questions it raises are the hardest to answer.'

What is GM? Can it go wrong?

Depending on who you talk to, genetic modification of crops is a panacea for our problems or a Pandora's box. Some argue that it is crucial in solving the food shortages that will no doubt result from the rising human population and the ravaging effects of climate change on the world. Detractors believe that playing with genes is an untested danger, and that scientists are using the world as their laboratory. Unlike in a lab, though, if things go wrong with their experiments, all of us will suffer.

Roger Beachy, head of the US National Institute of Food and Agriculture, is in the former camp. GM crops, he says, are an important tool in keeping farming sustainable and have already reduced the use of harmful pesticides and herbicides and the 'loss of soils because they promote no-till methods of farming. Nevertheless, there is much more that can be done.'

He points out that agriculture and forestry account for approximately 31 per cent of global greenhouse gas emissions, more than the 26 per cent from the energy sector. 'Agriculture is a major source of emissions of methane and nitrous oxides and is responsible for some of the pollution of waterways because of fertilizer run-off from fields. Agriculture needs to do better. We haven't reached the plateau of global population and may not until 2050 or 2060. In the interim, we must increase food production while reducing greenhouse gas emissions and soil erosion and decrease pollution of waterways. That's a formidable challenge. With new technologies in seeds and in crop production, it will be possible to reduce the use of chemical fertilizers and the amount of irrigation while maintaining high yields. Better seeds will help, as will improvements in agricultural practices.'

In the UK, similar arguments are bringing GM back to the political and scientific table, a decade after the 'Frankenstein foods' debacle, in which environmental groups successfully branded the technology irresponsible and harmful.

Today, GM technology is licensed for use in several parts of the world. The US is king when it comes to growing these crops, accounting for almost half of the world's production in 2009. Brazil is the second-biggest fan, at 16 per cent of the world's production in the same year.

In Uganda, farmers are trying out bananas modified (by the addition of a gene from green peppers) to resist a bacterial disease called banana Xanthomonas wilt, which has destroyed crops across central Africa and costs farmers half a billion dollars every year. The EU is more cautious. In 2010 it approved the cultivation of Amflora, a GM potato that contains a form of starch better suited for industrial use in making paper, adhesives and textiles. Before that, in 1998, officials had granted biotechnology company Monsanto permission to grow Mon 810 corn, which is resistant to the corn borer bug.

Release into the wild

In 2010, a genetically modified crop was found growing in the wild in the US for the first time. The plant, a type of rapeseed, was found in North Dakota, and according to some scientists highlighted a lack of proper monitoring and control of GM crops.

Cynthia Sagers, an ecologist at the University of Arkansas in Fayetteville, led the researchers who found two types of transgenic canola in the wild. One of them was resistant to Monsanto's Roundup herbicide (glyphosate), and one resistant to Bayer Crop Science's Liberty herbicide (gluphosinate). She also found plants that were resistant to both chemicals, showing that the GM plants had managed to interbreed. Even more intriguing, these wild populations did not exist just at the edges of large farms, but were found growing significant distances away.

In 2004, Swiss biotech company Syngenta announced that it had mistakenly labelled and sold the seed for an unapproved GM corn, Bt-10, as the GM corn approved for sale in the US, Bt-11. Both strains contain a gene from the soil bacterium *Bacillus thuringiensis*, which helps them to create their own pesticides against the corn borer. The Bt-10 strain is a laboratory version that is kept for research purposes, while Bt-11 is licensed for animal feed in the US.

Stories like this justifiably cause worry among consumers, though it is worth pointing out that there is little robust evidence so far that any GM crops are actually harmful when eaten.

In the UK, naturalists worry not only about the safety of the crops per se, but also about the broad-spectrum herbicides that come with them. These are so effective at killing everything other than the protected crop that they might strip out all the minor plants and seeds that farmland animals need to survive. When the population of tiny bugs and grubs in soil and plants falls, so do skylarks, partridges and corn buntings among others. In any farmland ecosystem, all the different forms of life compete for light and nutrients. With GM, the farmer has a formidable opportunity to outcompete everything else and use up all the resources for his crop alone.

In 2007, Emma Rosi-Marshall, an ecologist at Loyola University Chicago in Illinois, found that the larvae of caddis flies that fed on Bt corn debris grew only half as fast as those eating unmodified corn. In addition, caddis flies that ate pollen from Bt corn died at twice the rate of those feeding on normal pollen. Rosi-Marshall concluded that widespread planting of Bt corn might have 'negative effects' on local wildlife and perhaps 'unexpected ecosystem-scale consequences'.

Michael Antoniu, head of the nuclear biology group at Guy's Hospital in London and once an adviser to the UK government on GM foods, says that the problems with modification centre around the unintended consequences. 'It's a highly mutagenic process,' he told the *Observer* newspaper in 2008. 'It can cause changes in the genome that are not expected . . . These crops that have come along seem to be doing what they claimed they would be doing. The question is what else has been done to the structure of that plant? You might inadvertently generate toxic effects.'

Doug Parr, chief scientist at Greenpeace UK, says that if GM products were to get out into the environment, there would be no containing them. 'With the environment you could even create the problem simply by testing them.' The response of some nature groups has been direct action – ripping up plants from test fields.

And there might be a point to this behaviour, however unproven the science is. When genes from Bt corn were found in the wild plants of Mexico, experts were puzzled, because all genetically modified corn is illegal in that country. Corn is thought to have originated in Mexico, and the genetic bio-diversity there is important. If superweeds and superpests from biotechnology somehow managed to affect the wild plants, the genetic storehouse of this precious crop would be gone forever.

Biotech to order

Accidental leaks of GM crops into the environment are one thing, but there is not much evidence yet that they have done any damage. The technology that underlies GM, though, has been improving in recent decades, to the point where simple modification of DNA is not the only thing possible. Nowadays, scientists can create genes and life forms from scratch.

Craig Venter is the pioneer in this regard. In 2010, he created the world's first artificial organism, based on a bacterium that causes mastitis in goats. He produced genes from chemicals in the laboratory and inserted them into an existing bacterium, which used the synthetic genome to operate. According to Venter, the technology could be used to programme bacteria to make environmentally friendly biofuels or soak up harmful pollution from oil spills or the atmosphere.

Of course, it probably won't take long for someone to synthesize existing viruses or bacteria too, using publicly available genome information, or to design a new, rapidly expanding bug to which no human or important crop species has any immune resistance. Building and unleashing something like that on to an unsuspecting world would be catastrophic.

Safety first

The answer to the potential danger is not to ban or hide from the new genetic technologies. Regulation and oversight is key, but can it work?

'There is very little about the history of human activities involving living organisms that provides confidence that we can keep new life forms in their place,' says Arthur Caplan, a bioethicist at the University of Pennsylvania. He points out that for hundreds of years, people have been introducing new life forms into places where they create huge problems. 'Rabbits, kudzu, starlings, Japanese beetles, snakehead fish, smallpox, rabies and fruit flies are but a short sample of living things that have caused havoc for humanity simply by winding up in places we do not want them to be.'

When it comes to GM technology, scientists could build in failsafes – perhaps genes that prevent the crop from reproducing or which make them require particular compounds to grow properly. Unfortunately, this also leads to questions around control. When biotech company Monsanto did something like this with genetically modified seed containing 'terminator' genes, they were accused of enslaving poor farmers, who would have to buy new seeds every year from the multinational company.

Another idea, perhaps, is that synthetic life forms could be engineered to use a different amino acid code from natural organisms, so that they could be recognized in the wild and would be unable to reproduce with natural organisms.

Banning the technology outright, though, or applying more moratoriums is a non-starter. The solutions to any dangers posed by biotechology will come from biotechnology itself. And rather than being caught unawares, it has to be better to keep up to speed on information and ability when it comes to tackling any apocalyptic problems that might arise.

NANOTECH DISASTER

It is 2087, and there's been an oil spill off the coast of Alaska. A tanker carrying billions of gallons of crude oil has run aground, threatening the local environment with catastrophe. Fortunately, the authorities have a tried and tested weapon to fight the spill: a flotilla of tiny oil-munching robots that can break down hydrocarbons, rendering the spill harmless.

The machines, each one no wider than a human hair, can create more of themselves as they go along, so that there are always enough of them to deal with any size of oil spill.

But this time, something unexpected happens when the robots are dropped on to the spill. One of them has an error in its programming – instead of eating only hydrocarbons, this robot starts to eat anything with carbon in it. In other words, it sets itself to consume any living thing, along with its meal of oil. It doesn't take long before everything on Earth is consumed by the proliferating mass of robots. Life, as we know it, is gone.

Is is just a nightmare?

The end of the world brought about by self-replicating robots was an idea first put forward by Eric Drexler in his 1986 book *Engines of Creation*.

In the book, Drexler talked of the great possibilities and benefits of examining the world at the nano level. But he also warned of something more sinister. 'Plants' with 'leaves' no more efficient than today's solar cells, he said, could out-strip real plants, crowding the biosphere with an inedible foliage. 'Tough omnivorous "bacteria" could out-compete real bacteria: They could spread like blowing pollen, replicate swiftly, and reduce the biosphere to dust in a matter of days. Dangerous replicators could easily be too tough, small, and rapidly spreading to stop – at least if we make no prepara-tion. We have trouble enough controlling viruses and fruit flies.'

The result: a world turned into a featureless 'grey goo'; just a mass of these tiny robots rearranging atoms into copies of themselves. Science fiction authors picked up on the idea, notably in Michael Crichton's *Prey*, where nanobots run wild. Even Martin Rees, the UK Astronomer Royal and former president of the Royal Society, highlighted them as a potential cause of humankind's extinction.

Drexler's nightmare world would not take long to happen. 'Imagine such a replicator floating in a bottle of chemicals, making copies of itself … the first replicator assembles a copy in one thousand seconds, the two replicators then build two more in the next thousand seconds, the four build another four, and the eight build another eight,' he said. 'At the end of ten hours, there are not 36 new replicators, but over 68 billion. In less than a day, they would weigh a ton; in less than two days, they would outweigh the Earth; in another four hours, they would exceed the mass of the Sun and all the planets combined – if the bottle of chemicals hadn't run dry long before.'

What is nanotechnology?

Drexler's vision is worrying stuff. A 2003 editorial in the scientific journal *Nature* pondered whether nanotechnology was inherently dangerous, arguing that public calls for regulation of this rapidly developing and diverse discipline seemed to imply that it was. In response, the authorities seemed to think that nanotechnology could become a topic of public unease, 'and that the resulting debate will take place in an informational vacuum that will quickly be filled with hot air and hysteria'.

Later that year, the UK's Royal Society and Royal Academy of Engineering launched an investigation into the possible benefits and risks of nanotechnology, after Prince Charles had approached them with concerns about the technology.

When launching the Royal Society investigation, the UK's then science minister David Sainsbury acknowledged the difficulty of the task ahead: 'Nanotechnology could cover an enormous area. It is a bit like asking a committee when the first computer was designed to say: what is the impact of computers and IT going to be on the world in the future? The ability to predict far ahead is quite limited.'

Feynman's challenge

It is worth stepping back at this point and working out what nanotechnology really is. In December 1959, the great physicist Richard Feynman gave a lecture to the American Physical

Society titled 'There's Plenty of Room at the Bottom'. He talked about how to fit the entire twenty-four volumes of the *Encyclopaedia Britannica* on to the head of a pin, and calculated that it would be possible to write all the books in the world into a cube 1/200th of an inch wide, if they were encoded into strings of 1s and 0s, just as in computers. 'Computing machines are very large, they fill rooms. Why can't we make them very small, make them of little wires, little elements – and by little, I mean little.'

He finished his lecture with a challenge. 'It is my intention,' he said, 'to offer a prize of $1,000 to the first guy who makes an operating electric motor which is only 1/64th inch cubed.'

And this is how, in part, nanotechnology got its start. It began as a way of making computers and machines as small as possible, building them by rearranging individual atoms or molecules. One of the consequences of this drive for miniaturization is the fingernail-sized microchip in your computer, with wiring that is only eighty nanometres thick and which contains hundreds of millions of transistors.

Modern nanotechnology has blossomed into a mix of subjects that few could have predicted. It encompasses disparate fields, from medicine to space science to telecommunications, united only by scale. Researchers using the 'nano' prefix might all be working on objects that are a few millionths of a millimetre across, but they will all be doing very different things: fashioning materials never before seen in nature, tweaking molecular 'machines' found in bacteria or simply investigating the basic physics of what happens at really small scales.

Nanorobots, though, never survived as a real research endeavour after the initial interest in them. The closest anyone came to a miniature robot was the man who ended up winning

Feynman's $1,000 challenge, fellow Caltech scientist Bill McLellan. He spent five months in the early 1960s building a motor that was less than half a millimetre across, its wires only 1/80th of a millimetre wide, thinner than human hairs. It is not quite nanotechnology, mere microtechnology, and it burned out after a few uses.

What of the grey goo?

In 2004, Drexler made public attempts to play down his more apocalyptic warnings. 'I wish I had never used the term "grey goo",' he told *Nature*, adding that if he could write *Engines of Creation* again, he would barely mention self-replicating nanobots.

His statement underscored calculations made by a researcher at the Texas-based Zyvex Corporation, the first molecular nanotechnology company. Robert A. Freitas Jr looked at what it would take for Drexler's idea to come true. He weighed up how fast nanobots, if they ever came into existence, might be able to replicate, and how much energy they would have available to them against our ability to detect and stop them.

His research, published in 2000, concluded that fast-replicating devices of the type in Drexler's scenario would need so much energy and produce so much heat that they would become easily detectable to policing authorities, which could then deal with the threat.

If the nanomachines were made primarily of minerals containing aluminium, titanium or boron, then life forms would be spared the rampage anyway, as these metals are millions of times more abundant in the Earth's crust than in living things. The machines could just mine the Earth rather than killing us.

There is also the issue of power. 'Current nanomachine designs typically require power densities on the order of 10^5–10^9 W/m^3 (watts per metre cubed) to achieve effective results,' wrote Freitas. 'Biological systems typically operate at 10^2–10^6 W/m^3. Solar power is not readily available below the surface, and the mean geothermal heat flow is only 0.05 W/m^2 at the surface, just a tiny fraction of solar insolation.'

The Royal Society's report, published in 2004, also poured cold water on the idea that nanobots could replicate in the numbers required to destroy life. But the experts did raise several more pressing concerns about the possible health effects of the vanishingly small particles being made by the nanotechnology industry. They are created by grinding metals or other materials into an ultrafine powder – in sunscreens, for example, nanoparticles are designed to absorb and reflect UV rays while appearing transparent to the naked eye.

Ann Dowling, the Cambridge University professor who chaired the group behind the report, said: 'Where particles are concerned, size really does matter. Nanoparticles can behave quite differently from larger particles of the same material. There is evidence that at least some manufactured nano-particles are more toxic than the same chemical in its larger form, but mostly we just don't know. We don't know what their impact is on either humans or the environment.'

These particles, scientists warned, could be inhaled or absorbed through the skin. We already inhale millions of nanoparticles contained in the pollution frommotor vehicles, and these have been linked to heart and lung conditions. As nanotechnology becomes more widespread in industry, experts worry that we will become increasingly exposed to these airborne dangers.

Self-replicators

The idea of self-replication in nanotechnology has never quite gone away. But instead of building machines or replicators from scratch, as Drexler might have predicted in the 1980s, modern scientists look to nature for a helping hand.

Around the time Freitas wrote his paper, a researcher at Cornell University in New York state had attached a tiny nickel propeller on to a biological version of a motor powered by the fuel that makes humans tick, a molecule called ATP. At around ten nanometres across, this molecule is 50,000 times smaller than McLellan's mini motor. Adapting existing molecules like this, rather than building things from scratch atom by atom, is the way nanotechnology of the future will really work. And that is also where potential dangers lie.

'It is one of our most enduring myths that anyone with bad intentions will choose to express them in the most technologically complex way,' wrote journalist and author Philip Ball in an article for *Nature Materials*, bemoaning the scare tactics used by opponents of nanotechnology. 'You want life-threatening replicators? Then set loose a few smallpox viruses. These are nanobots that really work.'

ARTIFICIAL SUPERINTELLIGENCE

We have all come across the idea, in science fiction, of an ultra-smart computer or robot that tries to take over the world. Although it is designed by humans to improve their lives, its programming soon makes it recognize the superiority of its own abilities and technology over bog-standard flesh-and-blood life forms. Disaster awaits.

A familiar trope, perhaps, but not one that is too realistic for now, right? Modern computers do not approach the complexity or intelligence of even a human baby. Robots might be sophisticated enough to assemble cars and assist during complex surgery, but they are dumb automatons.

But don't discount progress. It is only a matter of time before the technological hurdles are surpassed and we reach a point where the machines are challenging us in the intelligence stakes. 'It might happen someday,' says Douglas Hofstadter, an expert in the computer modelling of mental processes at Indiana University, Bloomington. The ramifications would be huge, he says, since the highest form of sentient being on the planet would no longer be human. 'Perhaps these machines – our "children" – will be vaguely like us and will have a culture similar to ours, but most likely not. In that case, we humans may well go the way of the dinosaurs.'

The trouble with superintelligence

Intelligence is one of those very human qualities that is hard enough to define, never mind understand. Is it memory capacity? Is it processing ability? Is it the ability to infer meaning from multiple and conflicting sources of information simultaneously, in the way, for example, a teenager can when talking, texting, watching TV and surfing the web all at once?

The lack of a definition, however, has not stopped engineers and programmers from trying to recreate aspects of intelligence in machines. This endeavour has benefited all of us, giving us everything from cheap, fast computers to smart software that dynamically runs our electrical grids and traffic lights or manages our lives.

Compared to humans, though, even the most sophisticated modern machines are not what we would call 'intelligent'. There might be robots that can mimic basic emotions, and those that can have (almost) real conversations, but proper artificial intelligence is decades away at best. And probably even longer before it starts to become threatening.

Of course, another way of looking at it is that the arrival of true artificial intelligence is just a matter of time. Computing technology and robot controls roughly double in complexity and processing power every year. 'They are now barely at the lower range of vertebrate complexity, but should catch up with us within a half-century,' says Hans Moravec, one of the founders of the robotics department of Carnegie Mellon University. 'By 2050 I predict that there will be robots with humanlike mental power, with the ability to abstract and generalize. These intelligent machines will grow from us, learn our skills, share our goals and values, and can be viewed as children of our minds. Not only will these robots look after us in the home, but they will also carry out complex tasks that currently require human input, such as diagnosing illness and recommending a therapy or cure. They will be our heirs and will offer us the best chance we'll ever get for immortality by uploading ourselves into advanced robots.'

One way to build ever more intelligent machines is to keep copying humans. The human brain and nervous system is the most complex and intelligent structure we know of, and the technology to measure and copy it has taken off in the past few decades. 'In this short time, substantial progress has been made,' says Nick Bostrom, a philosopher and the director of the Future of Humanity Institute at the University of Oxford. 'We are beginning to understand early sensory processing. There are reasonably good computational models of primary visual cortex, and we are working our way up to the higher stages of visual cognition. We are uncovering what the basic learning algorithms are that govern how the strengths of synapses are modified by experience. The general architecture of our neuronal networks is being mapped out as we learn more about the interconnectivity between neurones and how different cortical areas project on to one another. While we are still far from understanding higher-level thinking, we are beginning to figure out how the individual components work and how they are hooked up.'

As well as raw processing power, researchers are adding more human-like properties: artificial consciousness, for example, could help machines understand their place in the world and behave in an appropriate manner, working out what is beneficial to it and its users, and what is dangerous.

Artificial emotions might also help us to have more natural and comfortable interactions with robots. Researchers at the University of Hertfordshire have programmed a robot to develop and display emotions, allowing it to form bonds with the people it meets depending on how it is treated. Cameras in the robot's eyes read the physical postures, gestures and movements of a person's body. So far, it can mimic the emotional skills of a one-year-old child, learning and interpreting specific cues from humans and responding accordingly. Its neural network can remember different faces, and this understanding, plus some basic rules about what is good and bad for it learned from exploring the local environment, allows the robot to indicate whether it is happy, sad or frightened about what is going on around it. It can also be programmed with different personalities – a more independent robot is less likely to call for human help when exploring a room, whereas a more needy and fearful robot will display distress if it finds something in the room that is potentially harmful or unknown.

The impact of artificial intelligence

If, as Moravec suggests, machines will achieve human-level intelligence by the middle of the 21st century, what are the implications? In an essay for the journal *Futures*, Bostrom plays out some of the consequences. First, he points out, artificial minds can easily be copied. 'Apart from hardware requirements, the marginal cost of creating an additional artificial intelligence after you have built the first one is close to zero.

Artificial minds could therefore quickly come to exist in great numbers, amplifying the impact of the initial breakthrough.'

As soon as machines reach human levels of intelligence, we will quickly see the creation of machines with even greater intellectual abilities that surpass any human mind. 'Within 14 years after human-level artificial intelligence is reached, there could be machines that think more than a hundred times more rapidly than humans do,' says Bostrom. 'In reality, progress could be even more rapid than that, because there would likely be parallel improvements in the efficiency of the software that these machines use. The interval during which the machines and humans are roughly matched will likely be brief. Shortly thereafter, humans will be unable to compete intellectually with artificial minds.'

This could be a good thing – these superintelligent machines will have easy access to huge sets of data, and will accelerate progress in technology and science faster than any human being. And they will no doubt also devote some of their energies to designing the next generation of machines, which will be even smarter. Some futurologists speculate that this positive-feedback loop could lead to what they call a singularity, a point where technological progress becomes so rapid that, according to Bostrom, 'genuine superintelligence, with abilities unfathomable to mere humans, is attained within a short time span'.

How do you predict whether such an intelligence would be a good thing for the human race? Would it help us to eradicate poverty and disease? Or would it decide that humans are a waste of resources and wipe us out? It could all be down to programming. 'When we create the first superintelligent entity, we might make a mistake and give it goals that lead it to annihilate humankind, assuming its enormous intellectual advantage gives it the power to do so,' says Bostrom. 'For example, we could mistakenly elevate a subgoal to the status of a supergoal. We tell it to solve a mathematical problem, and it complies by turning all the matter in the solar system into a giant calculating device, in the process killing the person who asked the question.'

Given the potential nightmare, it seems like a good idea to build in safeguards to prevent machines from hurting people or worse. 'The ethical question of any machine that is built has to be considered at the time you build the machine,' says Igor Aleksander, an emeritus professor of neural systems engineering at Imperial College, London. 'What's that machine going to be capable of doing? Under what conditions will it do it, under what conditions could it do harm?'

He adds that these are all engineering problems rather than ethical dilemmas. 'A properly functioning conscious machine is going to drive your car and it's going to drive it safely. It will be very pleased when it does that, it's going to be worried if it has an accident. If suddenly it decides, I'm going to kill my passenger and drive into a wall, that's a malfunction. Human beings can malfunction in that way. For human beings, you have the law to legislate, for machines you have engineering procedures.'

The science-fiction novelist Vernor Vinge considered what an Earth with a machine superintelligence might look like. 'If the singularity cannot be prevented or confined, just how bad could the Post-Human era be? Well . . . pretty bad.'

The physical extinction of the human race is one possibility, he says, but that is not the scariest scenario. 'Think of the different ways we relate to animals. Some of the crude physical abuses are implausible, yet . . . In a Post-Human world there would still be plenty of niches where human-equivalent automation would be desirable: embedded systems in autonomous devices, self-aware daemons in the lower functioning of larger sentients.'

He adds: 'Some of these human equivalents might be used for nothing more than digital signal processing. They would be more like whales than humans. Others might be very human-like, yet with a one-sidedness, a dedication that would put them in a mental hospital in our era. Though none of these creatures might be flesh-and-blood humans, they might be the closest things in the new environment to what we call human now.'

Is it likely?

You might be relieved to hear that not everyone is so pessimistic. In an interview on the potential dangers of a technology takeover, the linguist and psychologist Steven Pinker told the US Institute of Electrical and Electronics Engineers (IEEE) that there was not the 'slightest reason to believe in a coming singularity. The fact that you can visualize a future in your imagination is not evidence that it is likely or even possible. Look at domed cities, jet-pack commuting, underwater cities, mile-high buildings, and nuclear-powered automobiles – all staples of futuristic fantasies when I was a child that have never arrived. Sheer processing power is not a pixie dust that magically solves all your problems.'

But that does not mean it is impossible. John Casti of the International Institute for Applied Systems Analysis in Austria says that the singularity is scientifically plausible, and the only issue concerns the time frame over which it would unfold. This moment would mark the end of the supremacy of *Homo sapiens* as the dominant species on planet Earth. 'At that point a new species appears, and humans and machines will go their separate ways, not merge one with the other,' he told the IEEE. 'I do not believe this necessarily implies a malevolent machine takeover; rather, machines will become increasingly uninterested in human affairs just as we are uninterested in the affairs of ants or bees. But it's more likely than not in my view that the two species will comfortably and more or less peacefully coexist – unless human interests start to interfere with those of the machines.'

ENVIRONMENT

TRANSHUMANISM

Humans have spent thousands of years building tools to make our lives better, happier and more productive. At some point, our innovation will take us far beyond the natural and lead to people with capabilities so advanced that we might not define them as human. Are we ready?

One of the most important advances in the past century has been our increased understanding of medicine. Better drugs, better tools and a sophisticated knowledge of what happens to body cells when they go wrong – all of this has helped to heal us when we fall ill and gifted us longer, healthier lives.

Parallel to that in the past half-century has been the stupendous growth in information and digital technologies, resulting in our networked world. Today, everything you want to know is at your fingertips.

These twin tracks of knowledge intersect at many points – computing is crucial in basic biological research or in operating medical equipment – but could they be integrated further to make a better type of human? Future generations could have infinite memories thanks to computer implants. Their networked minds could bypass the fingertips to access all of the world's information. And as yet undreamed of medical breakthroughs will enable these improved humans to live for hundreds (perhaps thousands) of years. It all sounds so positive – what could possibly go wrong?

The transhumanist manifesto

Humanists believe that people matter; that even though we are not perfect as a species, things can be improved by freedom, tolerance, rational thinking and, above all, concern for other humans. People who call themselves 'transhumanists' agree with all of that, but also want to emphasize the potential that humans have to develop beyond our natural limits. 'Just as we use rational means to improve the human condition and the external world, we can also use such means to improve ourselves, the human organism,' says the philosopher Nick Bostrom, a proponent of the transhumanist way of thinking. 'In doing so, we are not limited to traditional humanistic methods, such as education and cultural development. We can also use technological means that will eventually enable us to move beyond what some would think of as "human".'

The future that Bostrom alludes to is known to trans-humanists as 'the era of the posthuman being'. This does not mean that humans will not exist any more; rather that the beings that will exist at that time will have basic capacities and desires so radically surpassing anything familiar to modern humans that they will be unrecognizable to us.

Transhumanists yearn to reach intellectual heights way above anything we know of today – as different from modern humans as humans are from other primates. They want un-limited youth and healthy life, and they want to be able to control their own desires and moods, so that they can avoid feeling tired or irritable, but can ramp up feelings of pleasure, love and artistic appreciation. They also want to experience states of consciousness inaccessible to modern human brains. 'It seems likely that the simple fact of living an indefinitely long, healthy, active life would take anyone to posthumanity if they went on accumulating memories, skills, and intelligence,' says Bostrom, who heads up the Future of Humanity Institute at the University of Oxford.

This posthuman world could end up being populated by very non-human-looking people. It could be a world of, say, artificial intelligences based on the thoughts and memories of humans who uploaded themselves into a computer and exist only as digital information on superfast computer networks. Their physical bodies might be gone, but they would be able to access and store endless information in milliseconds, and share their thoughts and feelings immediately and unambiguously with other digital humans.

A more recognizable posthuman might simply be a modern human who has been enhanced with genetic modification or drugs to slow ageing or mental decline. They might have neural interfaces, memory-enhancing prosthetics or be part cyborg to improve fitness and strength.

The rise of human modification

All of these scenarios might sound far-fetched, but technically, none of them is impossible. The accelerating rate of technological advance will lead us very soon into uncharted waters, and given the fact that most of our modern world was only developed in the past few decades, who knows where we will be in one or two centuries. Existing technologies, in their in-fancy now, already have the power to change our species.

Artificial intelligence, for example, is bound to happen in some form, allowing computers and robots to do the kind of thinking that has so far been the sole preserve of humans. No doubt the machines will eventually think faster and more creatively than humans – why not incorporate them into our physiology to make humans better? Brain–computer interfaces are already being trialled in patients with severe disabilities, allowing them to move computer cursors with the power of thought alone. Blind people have had electrodes fitted into their retinas to help them see for the first time in years.

The futurist Ray Kurzweil thinks that by 2035, human brains and computers will begin to merge. Tiny nanobots could be used to improve the way and the amount we can think, extending our intelligence. 'By 2020, $1,000 (£581) worth of computer will equal the processing power of the human brain,' he told the *Guardian* newspaper in 2005. 'By the late 2020s, we'll have reverse-engineered human brains . . . By 2030, we will have achieved machinery that equals and exceeds human intelligence but we're going to combine with these machines rather than just competing with them. These machines will be inserted into our bodies, via nanotechnology. They'll go inside our brains through the capillaries and enlarge human intelligence.' Kurzweil takes hundreds of dietary sup-plements each day and spends time every week at clinics having other health-giving compounds administered intra-venously – all this is intended specifically to keep him alive until such time when humans have worked out how to repro-gramme their own biology through nanotechnology.

What about more prosaic ideas of enhancement? In 2000,

the first draft sequence of the human genome was published by a consortium of international scientists. Researchers had been identifying and modifying genes for decades before that, but it had been a laborious and difficult process. The draft genome, combined with freefalling prices for sequencing technology, meant that the first decade of the 21st century was a glorious time for molecular biology, with an exponential growth in research that examined and manipulated genes.

At the same time, stem-cell biologists have been working on ways to understand disease and grow replacement tissue using the body's own master cells. In the future, reprogrammed stem cells will allow doctors to treat everything from heart failure to neurodegenerative diseases such as Parkinson's, Alzheimer's and diabetes.

Both genetic modification and stem-cell research are in their earliest stages, and are destined to be used at first in trials with people who are suffering from a disease. But transhumanists argue that the technology should not stop there – why not use it to enhance healthy people too, extending lifespan or cognitive ability?

The potential dangers

Technologies of all stripes tend to get cheaper as they become more entrenched, tested and commercialized. But for the first unknown number of decades, all the methods of enhancing human beings – stem cells, genetics, nanotechnology and computer intelligence – will no doubt be too costly for most people. This means that any advantages in lifespan or intelligence bestowed by these technologies will go disproportionately to the rich. This is nothing new – well-off people today already make more money and send their children to better schools – but it will only serve to exacerbate that divide.

Humans have always been at risk from each other, and to counteract this, we have created laws and institutions that act to prevent one group from suppressing another. But what if one group of humans was radically more capable than the rest? Laws and institutions would not worry them, or stop them from taking over the world, enslaving or killing all others.

Take uploads, for example, where a person manages to transfer their mind from their brain into a computer that can emulate all biological processes. 'Suppose uploads come before human-level artificial intelligence,' wrote Bostrom in an essay for the *Journal of Evolution and Technology*. 'A successful uploading process would preserve the original mind's memories, skills, values, and consciousness. Uploading a mind will make it much easier to enhance its intelligence, by running it faster, adding additional computational resources, or streamlining its architecture. One could imagine that enhancing an upload beyond a certain point will result in a positive feedback loop, where the enhanced upload is able to figure out ways of making itself even smarter; and the smarter successor version is in turn even better at designing an improved version of itself, and so on.'

If this runaway process happens quickly, it could result in one upload reaching superhuman levels of intelligence while everybody else remains at a normal human level. Such enormous intellectual superiority may give that person enormous power, allowing them to invent new technologies. If they were bent on domination, they might prevent others from getting the opportunity to upload. 'The posthuman world may then be a reflection of one particular egoistical upload's preferences (which in a worst-case scenario would be worse than worthless),' said Bostrom. 'Such a world may well be a realization of only a tiny part of what would have been possible and desirable.'

Future technology and our increasingly connected society could give very small groups of people, or even just a single person, the ability to control a huge amount from a small base. Today, our world and the people in it are separate enough that no single entity could do enough damage quickly enough to destroy all of it – if someone wanted to take over, at least some of us would survive. But what if the humans of the 27th century were all networked minds living mostly in computer networks?

The dangers are real, but, like the sharp sticks on that African plain thousands of years ago, progress should not be stopped unduly. 'You can't just relinquish these technologies,' says Kurzweil. 'And you can't ban them. It would deprive humanity of profound benefits and it wouldn't work. In fact it would make the dangers worse by driving the technologies underground, where they would be even less controlled.'

DEATH OF THE BEES

Could you live without fruit or vegetables? What about clothes made from cotton? How would you feel if the world's meadows had no flowers?

All of these plants, so crucial to the continuing economic and aesthetic success of our lives, survive through the generations thanks to insects that are attracted by their flowers and their smells. They come to eat nutritious nectar and inadvertently carry pollen from one plant to another. A third of everything we eat depends upon pollination by bees, moths and hoverflies, which means that these creatures contribute some $42 billion to the global economy. If all of the UK's insect pollinators were wiped out, the drop in crop production would cost the economy up to £440 million a year, equivalent to around 13 per cent of the country's income from farming.

The king of this world is the bee. There are hundreds of species of bees, all of them critical in the life cycle of the various plants around the world, ranging from apples, carrots, oranges and onions to broccoli, melons, strawberries, peaches and avocados.

But there is something amiss: bees are disappearing, and fast. Since the problem was first identified in Britain in the

1950s, numerous studies have documented long-term deterioration in bee species. In fact, all pollinating insects have been in serious decline around the world in the latter half of the 20th century, a result of disease, changing habitats around cities, and increasing use of pesticides.

A study by Sydney Cameron, an entomologist at the University of Illinois, illustrates how fast it is happening. She looked at the genetic diversity and pathogens in eight species of bumblebees in the US. Her results, published in 2011 in the *Proceedings of the National Academy of Sciences*, showed that numbers of four common species of bumblebee in the US have collapsed by 96 per cent in just the past few decades.

By comparing modern census data about the insects with those in museum records, she also found that the geographical spread of four of the bee populations she studied (*Bombus occidentalis, B. pennsylvanicus, B. affinis* and *B. terricola*) had contracted by between 23 per cent and 87 per cent, some within the past two decades.

The findings reflected those of similar studies across the world. According to the Centre for Ecology and Hydrology in the UK, three of the twenty-five British species of bumblebee are already extinct, and half of the remainder have shown a serious fall in number, often up to 70 per cent, since around the 1970s. In addition, around 75 per cent of all butterfly species in the UK have been shown to be in decline.

Canada, Brazil, India and China, as well as most of western Europe, have similar bee afflictions. The US National Research Council warns that bees could be extinct in North America by 2035.

The insect economy

Bumblebees are important pollinators of wild plants and agricultural crops around the world thanks to their large body size, long tongues and high-frequency buzzing, which helps release pollen from flowers.

Bees in general pollinate some 90 per cent of the world's commercial plants, including most fruits, vegetables and nuts. Coffee, soya beans and cotton are all dependent on pollination by bees to increase yields. It is also the start of a food chain that sustains wild birds and animals.

Pollinators are crucial for the quality of fruits and vegetables. Perfectly shaped strawberries, for example, are created only if every single ovary has been pollinated by an insect. And the number of seeds in a pumpkin depends on the number of species of insect that have pollinated the plants. 'If you've got ten pollinators, you'll get more seeds in the pumpkin than you would have got if you've just got one pollinator,' says Giles Budge of the UK Food and Environment Research Agency. 'It is important to have that diversity in a pollinating population.'

Causes of decline

So what's causing the crash? Scientists think it is a combination of disease and changing agricultural practices. 'The spread of industrial farming, increased use of pesticides, and loss of habitat led to declines in the role of wild insect populations such that they are now reported to account for just 15% of global crop pollination,' says Elliott Cannell, coordinator of the Pesticides Action Network, Europe. 'In response farmers started to hire in honeybees to pollinate their fields, thus creating a market for pollination. Demand soon spawned an industry which today sees honeybees overexploited, plagued by parasites, exposed to pesticides, and ill adapted to the conditions they work in.'

In her study on American bees, Sydney Cameron pointed to two causes: a pathogen called *Nosema bombi,* and an overall reduction in the genetic diversity of the remaining bee populations. The pathogen, common in bumblebees throughout Europe, reduces the lifespans of individual bees and also results in smaller colony sizes. Reduced genetic diversity means that the smaller populations are less able to fight off any new pathogens or resist pollution or predators.

Another problem for bees is the blood-sucking varroa mite. This creature has been endemic in the honeybee populations of Asia for thousands of years, living symbiotically with the local *Apis cerana* population. The completion of the Trans-Siberian Railway in 1916 and the movement of trade and people along its length, however, kicked off an unintended problem for the Western honeybees. These bees had never been exposed to the mite and therefore had no natural methods of protection against them.

By the 1950s, varroa had entered the Soviet Union, and two decades later had spread to eastern Europe and South America, thanks to the movement of bee populations by people. Today, Australia is the only continent free of varroa. Billions of honeybees around the world have died as a result of the mite, which spreads viruses deadly to the insects.

But virus epidemics alone are not enough to explain the massive insect decline. By far the biggest danger to bees in the past few decades has been our increased use of pesticides, specifically those called neonicotinoids.

One winter in the early 1990s, French beekeepers noticed a sudden fall in their insect population. They quickly pointed to a best-selling pesticide called imidacloprid that had been used for the first time the year before. The US Environmental Protection Agency classes it as 'highly toxic' to honeybees. Following the death of so many bees in one go, France banned imidacloprid.

In 2008, Germany suspended three neonicotinoid pesticides after reports from beekeepers in the Baden-Württemberg region that two thirds of their bees had died around places that had used a pesticide called clothianidin. The chemical had been applied to the seeds of sweetcorn planted along the Rhine.

The problem is widespread. A survey carried out in 2008 by researchers at Pennsylvania State University showed evidence of seventy pesticides or breakdown products in pollen and bees. All the bees they looked at had traces of at least one pesticide, while each pollen sample had around six pesticides, with as many as thirty-one in one case.

Research on the effects of pesticides shows that the chemicals damage the brains of the bees, blocking the electrical and chemical signals between neurons. According to bee experts, only subtle changes would be required to produce serious brain disorders in the insects. The results would include making it harder for the insects to get back to their hives after foraging trips; or interfering with their ability to communicate with nest-mates using the 'waggle dance', where bees come back to their hive and spread information about the food sources they have found.

Can we stop it?

Monitoring and further research into the causes of bee decline will be the first step. In November 2010, scientists meeting at the Saint Louis Zoo in Missouri recommended that due to their low numbers, *B. affinis*, *B. terricola* and *B. occidentalis* should be listed on the Red List of Threatened Species compiled by the International Union for Conservation of Nature (IUCN). They also proposed the creation of a bumblebee specialist group within IUCN to help policymakers and governments combat the population drops in bees.

Scientists at the University of Bristol are working on identifying hot spots of insect biodiversity in Bristol, Reading, Leeds and Edinburgh, in a bid to make cities more friendly to bees and other insects. They want to map everything from gardens to bits of wasteland, industrial estates and shopping centres to work out where there might be potential oases for pollinating insects.

Others are analysing DNA from live wild bees in order to track how far and wide queen bees fly to start new nests, and how far worker bees go to look for food. Conserving populations means thinking about the number of nests and not just the number of individual bees – though there is a big challenge in that it is almost impossible to find bumblebee nests in the wild.

If the conservation efforts do not stem the tide, perhaps we will need to start looking for new pollinators to replace bumblebees. The US Department of Agriculture's pollinating insect research unit at Utah State University has been studying the blue orchard bee.

Like honeybees, blue orchards can pollinate a variety of plants, including almond, peach and apple trees. They do not live in hives, however, preferring to spend their time in bore-holes made in dead trees by other creatures, or in holes drilled into pieces of wood by people. And they are very effective: 2,000 blue orchards can do the work of 100,000 honeybees when it comes to pollinating blossoms on fruit trees.

Wherever the future lies, let's hope the buzzing of the bees never goes away.

INVASIVE SPECIES

The Earth's plants, animals and microorganisms have evolved together through a complex web of interactions over billions of years. Each of the world's ecosystems is carefully balanced to keep its specific inhabitants alive. Insert something from outside, however, a plant or animal that nature never intended to be there, and all hell can break loose.

Individuals of a species compete for food and space, animals hunt or hide from each other, microbes can be symbiotic or parasitic with other plants or animals. Over time, this interplay has created balanced ecosystems of diverse life all around the world, where different species can coexist without huge drama or sudden upsets.

Imagine what happens, then, if an invader encroaches on a balanced ecosystem. That invader could be a plant or animal that has not shared the co-evolution of the other myriad life forms already present in the ecosystem. If it were a predator animal, an aggressively growing plant or a pathogenic microbe, it would find easy pickings in a place where none of the species it encountered had evolved defences against it. With no way of keeping the predatory behaviour in check, the invader's population would surge, and the once-balanced ecosystem would quickly destabilize towards collapse.

'Ask an Asian rice farmer about a brown or green-coloured snail, some 10 cm in length, and you could well be asking about sinister creatures from Mars,' says Achim Steiner, executive director of the United Nations Environment Programme. 'The golden apple snail has become a scourge in the paddy fields, damaging a staple crop as a result of its voracious appetite and costing a small fortune to control via environmentally questionable chemicals. The mollusc is among literally tens of thousands of life-forms classed as alien invasive species.'

Invasive species cost the global economy at least $1.4 trillion per year. They disrupt and destroy local ecosystems, transfer viruses around the world, poison soils and damage agriculture. And there is no stopping them.

What are invasive species?

The idea of invading creatures might bring to mind biblical plagues of locusts. Or perhaps the aliens that arrived to systematically destroy humans in H.G. Wells' *War of the Worlds*. Each scenario is terrifying for its own reasons, but neither of them comes close to the current devastation caused as humans inadvertently spread species around the world and force meetings between organisms that were never meant to meet naturally.

In part, this is one of the side effects of our globalized world. More than 90 per cent of trade is carried by sea, and the volume of traffic is set to double by 2018, with the global fleet increasing by 25 per cent. In addition, hundreds of millions of sea passengers pass through European ports every year.

Increasing tourism and trade allows species to hitchhike between countries or continents on travellers' baggage or clothes, or in freight or the ballast or waste water of ships. Plants and animals cling to the sides of ships in one country's waters, ending up a few months later in the harbours and waterways of a country thousands of miles away. When fish started dying in their droves in the North Sea in the 1990s and 2000s, the deaths were linked to blooms of algae brought in by accident in the ballast water from the seas off China.

The golden apple snail was brought to Asia from South America in the 1980s as an aquarium pet and as gourmet food. When the snails did not sell, the importers released them into Asia's lakes, and they have now spread to a dozen countries.

Travelling the other way around the world is the Asian long-horned beetle, which is normally endemic in Japan, Korea and China. The females of this species chew holes in the bark of hardwood trees (such as ash, maple, chestnut and willow) to lay eggs. When these hatch, the larvae bore further into the trunk and the damage they cause can kill the tree.

In recent decades, the beetles have been found in North America and Europe, imported via wood or wooden packaging. An infestation in northern Italy, caused by a single adult on four trees, required the felling of more than 300 susceptible trees in the vicinity in order to quarantine the pest.

Around ten new species become established every year in Europe, and there is a rising trend for invertebrates and fish to be introduced. According to a study on non-native species for the UK government, carried out by the environmental group CABI, the average time it takes for a species to get rooted in a new habitat is about fifty years, but this period is shorter in tropical species than in species from temperate regions. 'In general, however, the rate of spread of [invasive species] is often exponential,' says the report.

Effects on biodiversity and ecosystem services

There are more than a billion hectares of hardwood forest in Great Britain. The government has estimated that if the Asian long-horned beetle became established in these forests, it would cost the economy more than £430 million. This includes the loss of income from infested trees as well as trees that have to be cut down to quarantine an area. For a country the size of the US, this figure jumps to a staggering $138 billion if the hardwood industry has to pack up because of beetle infestation.

Invasive plants can also destroy the way people live. 'Take water hyacinth as one example,' says Steiner. 'A native of the Amazon basin, it was brought to continents like Africa to decorate ornamental ponds with its attractive violet flowers. But there is nothing attractive about its impacts on Lake Victoria, where it is thought to have arrived in about 1990, travelling down the Kigera River from Rwanda and Burundi.'

Floating blankets of hyacinth, continues Steiner, have affected shipping, reduced fish catches, hampered electricity generation and had an effect on human health. 'The plant has now invaded more than 50 countries around the world and annual costs to the Ugandan economy alone may be $112m. In sub-Saharan Africa, the invasive witchweed is responsible for annual maize losses amounting to $7bn. Overall losses to aliens may amount to more than $12bn in respect to Africa's eight principle crops.'

INVASIVE SPECIES

Invasive species are acknowledged as one of the main pressures driving the loss of biodiversity around the world. 'Of the 174 European species listed as critically endangered by the IUCN Red List, 65 are in danger because of introduced species,' says a 2010 report by the Institute for European Environmental Policy (IEEP). These include some of the most threatened species, such as the European mink and the ruddy duck. 'The cumulative number of alien species is increasing for all groups including mammals, with one new alien mammal introduced per year. Similar patterns are observed in Europe's marine environment.'

At the global level, says the IEEP report, invasive species have been identified as a key factor in 54 per cent of all known extinctions documented by the Red List of endangered species, and the only factor in 20 per cent of extinctions. 'They are the second most important pressure on birds, impacting over half of species listed as critically endangered, the third most severe threat to mammals and the fourth to amphibians.'

Invasive species can also pose a health risk for humans. Cockroaches, for example, can carry pathogenic bacteria on to food and cause poisoning. 'A survey by the Chartered Institute of Environmental Health in the Cossall Estate in London, which consists of 421 apartments contained in eight three-storey blocks with a history of cockroach infestation, found that 15.7 per cent of the apartments were infested. An estimated 59 per cent of English hospitals were reported to have cockroach infestations,' says the CABI report for the UK government.

Economic costs

Invasive plants and animals can destroy local biodiversity and disrupt the food web. But they can also ruin agricultural crops and compromise key ecosystem services, such as pollination or keeping water clean, needed to sustain life or make money from the land.

In the Philippines, the golden apple snail is responsible for damage of up to $45 million to the annual rice crop. The IEEP report says that lost output due to invasive species, health impacts and expenditure to repair damage has already cost Europe at least 12 billion euros per year over the past twenty years.

More than $3.5 billion of crops are lost in the UK every year because of invasive pests and weeds. In the US, major environmental damage and losses from the 50,000 or so invasive species added up to almost $120 billion per year, according to a 2004 study by David Pimentel of the College of Agriculture and Life Sciences at Cornell University.

What can we do?

Some countries recognize the potential danger of invasive species and have tough rules on bringing foreign plants and animals across their borders. But according to Steiner, not enough places have really grasped the scale of the threat. The solution, he says, is to boost the capacity of customs, quarantine and scientific institutes able to provide early warning of invasive species, and this is particularly true for the world's developing countries.

'Improved management of affected habitats can also assist,'

he concludes. 'There is some evidence that introducing a variety of native freshwater plants into a golden apple snail-infested site can reduce impacts on the rice crop . . . As the economy recovers, global trade, including via shipping, will resume the risk of further invasions. Alien invasive species are part of the overall biodiversity challenge; for too long they have been given an easy ride.'

DESERT EARTH

It has been called 'the greatest environmental challenge of our time' and 'a threat to global wellbeing' by a top UN official. It will displace millions, if not billions of people from their homes, cause wars and prevent people from growing the food they need to survive.

The slow degradation of large swathes of the Earth into inhospitable desert is a natural process that occurs over millions of years as the climate shifts. But now, thanks to our treatment of the environment and the increase in greenhouse gases in the atmosphere, we are speeding this process up so that it is happening in the lifetimes of individuals, in parts of the world that can barely afford to deal with the consequences.

'The top 20 cm of soil is all that stands between us and extinction,' says Luc Gnacadja, executive secretary of the UN's Convention to Combat Desertification (UNCCD). Healthy soil locks in a huge amount of carbon, and the organisms living in it are crucial for the growth of crops and forests. Gnacadja blames degradation and overuse of the land for conflicts seen in Somalia, dust storms in Asia and the increases in the price of food in recent years. A quarter of the planet's land has been rendered useless since the 1980s, and this process continues at the rate of 1 per cent per year today.

According to the UN's food and agricultural organization, 75 billion tonnes of soil, which equates to around 10 million hectares of arable land, is lost every year to erosion, waterlogging and salination. A further 20 million hectares is abandoned because its soil quality has been degraded.

And this is a problem that cannot be fixed quickly. Lester Brown of the Earth Policy Institute says that it takes up to 1,000 years to renew a layer of soil an inch thick. 'The thin layer of topsoil that covers the planet's land surface is the foundation of civilisation,' he says. 'This soil, typically 6 inches or so deep, was formed over long stretches of geological time as new soil formation exceeded the natural rate of erosion. But sometime within the last century, as human and livestock populations expanded, soil erosion began to exceed new soil formation over large areas.'

With world populations on the increase, the pressure on land to produce ever more food will keep on rising. 'Increased aridity is making the drylands the most conflict prone region of the world,' says Gnacadja. 'If you really want to look at the root causes of the conflicts in Somalia and Darfur, and the drylands of Asia, you will understand that people in their quest to have access to productive land and water for life, they end up in conflict.'

How are deserts formed?

The great expanse of desert that covers a third of the Earth's surface was formed by interactions between land and climate over millions of years. These drylands are home to more than

2 billion people, and the area has grown and shrunk as vegetation and rainfall have come and gone and humans have used or misused the land. 'These arid regions are called deserts because they are dry,' according to the US Geological Survey (USGS). 'They may be hot, they may be cold. They may be regions of sand or vast areas of rocks and gravel peppered with occasional plants. But deserts are always dry.'

Desertification is the process of degradation of land that has once been productive, as a farm or in supporting wildlife and forests. Deserts themselves grow and shrink naturally, and 'areas far from natural deserts can degrade quickly to barren soil, rock, or sand through poor land management', says the USGS. 'The presence of a nearby desert has no direct relationship to desertification. Unfortunately, an area undergoing desertification is brought to public attention only after the process is well underway. Often little or no data are available to indicate the previous state of the ecosystem or the rate of degradation.'

Perhaps the best-known desertification in recent times occurred in the 1930s, when the Great Plains of the US turned into the 'Dust Bowl' due to drought and overgrazing by livestock. More than 3.5 million people were forced to abandon their homes and livelihoods during this time. Nowadays, dust storms from the Gobi Desert in China blow into Beijing and surrounding countries, including South Korea. The biggest storms even reach North America.

When we try to remove too many resources – this includes excessive farming, diversion of water, grazing by livestock and building human settlements – the land will become unsustainable. 'Increased population and livestock pressure on marginal lands has accelerated desertification,' says the USGS. 'In some areas, nomads moving to less arid areas disrupt the local eco-system and increase the rate of erosion of the land. Nomads are trying to escape the desert, but because of their land-use practices, they are bringing the desert with them.'

It is also a misconception that droughts are at the root of desertification. Sure, dry conditions are common in at-risk places, but well-managed lands can recover from drought when the rains return, according to the USGS. 'Continued land abuse during droughts, however, increases land degrada-tion. By 1973, the drought that began in 1968 in the Sahel of West Africa and the land-use practices there had caused the deaths of more than 100,000 people and 12 million cattle, as well as the disruption of social organizations from villages to the national level.'

When land has been compromised and becomes desertified, it can no longer support a wide array of plants and animals and it becomes virtually useless to life. Deserts destroy bio-diversity and prevent the growth of crops, leading to starvation. In 2008, the degradation of available farmland led to global spikes of 130 per cent in the cost of wheat over the previous year and a rise of more than 80 per cent in the cost of soy.

In Africa, almost 75 per cent of the land is classed as dry and is affected by potential desertification due to the rise in farming – in 1950, the continent was home to 227 million people and 273 million livestock; just over half a century later, there are almost a billion people and more than 800 million livestock.

But these days, the great frontier in unchecked desertifi-cation is in China. More than a quarter of the country's land is either degraded or lost to sand and gravel – a combined result of the dry climate, hundreds of years of overcultivation, and excessive demand on water and soil as the economy has grown faster than any other in the world.

The amount of livestock in China has increased rapidly, but the land available for grazing has decreased. Scientists there report that between 1950 and 1975, around 1,550 km^2 (600 sq miles) of land turned to desert every year. By the turn of the century, that had jumped to 3,625 km^2 (1,400 sq miles) per year. 'Over the last half-century, some 24,000 villages in northern and western China have been entirely or partly abandoned as a result of being overrun by drifting sand,' says Lester Brown.

Can we stop it?

Stopping or slowing desertification and its impacts requires farmers to adopt better land management practices, and use their water more sustainably. They could also plant drought-resistant seeds in stressed areas – a good way to provide food and keep the soil together.

Alternatively, physical boundaries could be employed. The Chinese authorities have planted a wall of trees, stretching more than 4,830 km (3,000 miles) in length, from outer Beijing to Mongolia, in an attempt to protect their cities from the dust storms of the Gobi. In 2011, the planting of an 8,000 km (5,000 mile) long, 14.5 km-(9 mile) wide wall of trees from Djibouti in the Horn of Africa in the east to Dakar in Senegal was approved during a meeting of the UNCCD – it is hoped that this 'pan-African Great Green Wall' will stem the desertification of the Sahel.

Modern farming techniques might also help. Traditional farming involves ploughing up land and then dropping seeds on to it. This has worked for generations, but it disrupts the structure of the soil and allows more of it to erode away during rain or wind storms. 'No-till' agriculture has helped to address the problem somewhat: in this technique, farmers drill seeds straight into the ground. This method has taken off in many countries, including the US, Brazil, Argentina and Canada.

Covering sand dunes with large boulders or petroleum can also disrupt the flow of winds and prevent the dust from moving. Placing straw grids on the surface can reduce the speed of the wind. 'Shrubs and trees planted within the grids are protected by the straw until they take root,' says the USGS. 'In areas where some water is available for irrigation, shrubs planted on the lower one-third of a dune's windward side will stabilize the dune. This vegetation decreases the wind velocity near the base of the dune and prevents much of the sand from moving. Higher velocity winds at the top of the dune level it off and trees can be planted atop these flattened surfaces.'

There are even more ambitious plans to re-green deserts. The Sahara Forest project would marry huge 'seawater' green-houses with concentrated solar power (CSP), which uses mirrors to focus the sun's rays and generate heat and electricity. The installations would turn deserts into lush patches of vege-tation, according to the project's designers, without the need to dig wells for fresh water. The scheme works by using the solar power to run seawater evaporators and then pump the damp, cool air through the greenhouses. This reduces the tem-perature by about 15°C compared to that outside. The water vapour is condensed at the other end of the greenhouse from the evaporators. Some of this fresh water is then used to irri-gate the crops, while the rest can be employed for the essential task of cleaning the solar mirrors. In test greenhouses, the crops sitting in these slightly steamy, humid conditions grew fantastically well.

The Chinese authorities are planning to spend billions on anti-desertification measures – including tree-planting, the relocation of millions of people and restrictions on herding and farming – but they admit it could take 300 years to make a significant impact.

We have the solutions to stop the degradation of the planet. There seems to be some political will too. The big question is whether we can hold our nerve for the generations it will take to fix the problem.

GLOBAL FOOD CRISIS

What happens when the price of bread goes up in London, or the availability of tortillas drops in Mexico? If food becomes so expensive that only the richest can eat, what will happen to the countries most in need of development? Will they starve? If agriculture failed us, would the world survive?

Agriculture is one of the things that makes civilization what it is today. And technological improvements have kept it relevant, feeding ever more people, ever more efficiently. But our demands on it will keep increasing as the number of people in the world goes up and our changing climate puts strain on the areas of the world available to be farmed for food. Can we keep up?

We've been here before, haven't we?

Scenarios of food shortage have been around for centuries. In 1798, the English cleric Thomas Malthus warned of the perils of unchecked human reproduction – the challenge of feeding so many people, he said, would bring us to the limits of agriculture, and human lives would be lost as a result. His warnings never materialized, because of the industrialization of farming and the introduction of new technologies that used land more efficiently and moved food around the world more quickly. Fertilizer made crops grow bigger, irrigation transformed barren land into farmland, specially selected varieties of wheat grew ever more productively.

Every time a dusty academic foretold the end of human civilization because of the increased pressure on the planet's potential food resources, they were somehow proved wrong. So what is to say that current warnings have any merit?

Lester Brown, arch-environmentalist and founder of the Worldwatch Institute and the Earth Policy Institute, sums up many of the concerns about 21st-century food shortages as a major threat to global stability.

According to Brown, the 'green revolution' in the 1960s and 1970s that used technology fixes to boost agricultural productivity has hit its limits, and the long-term rise in what we can grow on a piece of land is slowing down. 'Between 1950 and 1990 the world's farmers increased the grain yield per acre by more than 2% a year, exceeding the growth of population,' said Brown in an article for *Scientific American* in 2009. 'But since then, the annual growth in yield has slowed to slightly more than 1%. In some countries the yields appear to be near their practical limits, including rice yields in Japan and China.'

One set of factors concerns supply – our environment is coming close to the limit of what it can produce. Temperature rises thanks to climate change are a real threat to global crops: the US National Academy of Sciences reckons that for every 1°C rise in surface temperatures, wheat, rice and corn yields drop by 10 per cent. Add to that the world's rapid overuse of fresh water, which has led some countries to mine fossil aquifers and other non-renewable sources to feed their crops. And topsoil is eroding too, faster than new soil can form. 'This thin layer of essential plant nutrients, the very foundation of civilization, took long stretches of geologic time to build up, yet it is typically only about six inches deep,' says Brown. 'Its loss from wind and water erosion doomed earlier civilizations.'

On the other side, demand for grain is rising. This is due partly to rising affluence. 'People in low-income countries where grain supplies 60% of calories, such as India, directly consume a bit more than a pound of grain a day,' says Brown. 'In affluent countries such as the US and Canada, grain consumption per person is nearly four times that much, though perhaps 90% of it is consumed indirectly as meat, milk and eggs from grain-fed animals.'

As increasing numbers of people in India and China start to consume meat, because more of them can afford something that was once too expensive, the demand for grain will become that much more urgent.

Kenneth Cassman, an agronomist at the University of Nebraska–Lincoln, believes that humanity has enjoyed an unusual streak of food surplus since the green revolution began in the mid-1960s. 'These trends sustained economic development and a significant reduction in global hunger and poverty,' he wrote. 'A sharp reversal is now possible, however, given strong economic growth in the world's most populous countries and loss of suitable cropland. People with rising incomes consume more meat and livestock products, which in turn requires more grain per unit of food produced. The rapid expansion of biofuel production only complicates the competition between food and fuel.'

Around the world, governments are championing the production of biofuels to replace traditional fossil fuels for cars and lorries, part of their acknowledgement that we need to do something to prevent climate change. Farmers are being incentivized to grow maize, sugar cane, palm oil and oilseed rape, all of which can be turned into ethanol or other sustainable fuels.

Figures from the US Department of Agriculture show that a quarter of the maize and other grain crops grown in that country in 2008 was used to create ethanol. At average world consumption levels, that 107 million tonnes of grain could instead have fed 330 million people for a year. And the problem is set to get worse – President George W. Bush challenged farmers to increase their production of biofuels by 500 per cent by 2017, to 35 billion gallons per year, to help cut down on foreign oil imports. 'The US, in a misguided effort to reduce its dependence on foreign oil by substituting grain-based fuels, is generating global food insecurity on a scale not seen before,' said Brown.

As if to underline the point, a World Bank report in 2008 concluded that the drive for biofuels by American and European governments had pushed up food prices by 75 per cent. Loaves of bread get more costly in England, tortillas become unaffordable in Mexico. Rising food prices sent thousands of demonstrators on to the streets across countries in the southern Mediterranean and north Africa.

How food shortages bring about a failed state

If costlier food was the only result of the looming shortage, it might stay as a development issue, one of many. But the crisis will have repercussions down the chain to wider society. 'Food scarcity and the resulting higher food prices are pushing poor countries into chaos,' says Brown. 'Such "failed states" can export disease, terrorism, illicit drugs, weapons and refugees. Water shortages, soil losses and rising temperatures from global warming are placing severe limits on food production. Without massive and rapid intervention to address these three environmental factors . . . a series of government collapses could threaten the world order.'

In a failing state, the government can no longer provide personal security or basic social services such as education and health care. Food security is out of the window.

'For most of us, the idea that civilization itself could disintegrate probably seems preposterous,' wrote Brown. 'Who would not find it hard to think seriously about such a complete departure from what we expect of ordinary life? For many years I have studied global agricultural, population, environmental and economic trends and their interactions. The combined effects of those trends and the political tensions they generate point to the breakdown of governments and societies. Yet I, too, have resisted the idea that food shortages could bring down not only individual governments but also our global civilization. I can no longer ignore that risk. Our continuing failure to deal with the environmental declines that are undermining the world food economy – most important, falling water tables, eroding soils and rising temperatures – forces me to conclude that such a collapse is possible.'

If you think a shortage of food will not affect affluent people in the West, think again. Failing states have global impacts as a source of terrorists, drugs, weapons and refugees. Citizens of these states can threaten political stability in their own region and much further afield too. Among the failing states identified by the United Nations, says Brown, Somalia has become a base for piracy, Iraq is a hotbed of terrorist training and Afghanistan is the world's leading supplier of heroin.

Rising food prices spread disorder. In Thailand, rice thieves have forced villagers to guard their crop at night with shotguns. Trucks carrying grain are hijacked in Sudan.

If individual nation states start to break down, international finance becomes more difficult, and the global spread of disease is harder to manage. 'If the system for controlling infectious diseases – such as polio, SARS or avian flu – breaks down, humanity will be in trouble,' says Brown. 'If enough states disintegrate, their fall will threaten the stability of global civilization itself.'

What can we do?

There may be technological fixes on the horizon. Crops might be genetically modified to produce higher yields or to grow on land that has little water or nutrients in its soil. But no such crops have been tested on a large enough scale to call this idea anything but fantasy for now.

Perhaps, though, the solutions are closer to home. Instead of relying on technology, we might think about altering our behaviour. Conserve soils, for example, or grow and eat wheat rather than rice (the latter uses up far more water).

A 2010 editorial in *Nature* was optimistic about our chances. Producing enough food for the world's population would be easy, it said, though doing it sustainably would be difficult. 'Clearing hundreds of millions of hectares of wildlands – most of the land that would be brought into use is in Latin America and Africa – while increasing today's brand of resource-intensive, environmentally destructive agriculture is a poor option. Therein lies the real challenge in the coming decades: how to expand agricultural output massively without increasing by much the amount of land used.'

The journal advocated a new position: a second green revolution described by the UK's Royal Society as the 'sustainable intensification of global agriculture'. That means finding new crop varieties that use less water and are more resistant to pests and heat, as well as curbing waste everywhere in the system – up to a third of the food produced in the world is lost or spoiled.

'Meeting these goals may be necessary to prevent the collapse of our civilization,' says Brown. 'Yet the cost we project for saving civilization would amount to less than $200 billion a year, a sixth of current global military spending. In effect, [it] is the new security budget.'

WATER WARS

How on Earth could we run out of water? It sounds absurd to even think it, given how much of our planet is covered in the stuff. Water, that essential ingredient for life and all the things that go with it (including food, power generation and manufacturing), is everywhere. Right?

In total, there are some 33 million km^3 (8 million cubic miles) of fresh water in the world – many thousands of times the amount humans consume every year. Talking averages, people drink around a cubic metre of water a year and use one hundred times that amount for cleaning and washing. Growing the food we eat annually takes up another 1,000 cubic metres.

So why is it that global governments and world bodies such as the Intergovernmental Panel on Climate Change (IPCC) and the United Nations keep warning about a coming water apocalypse? Or potential skirmishes over access to water resources? Or billions of people who by the middle of this century will be without clean water?

It seems counterintuitive, for sure, but there is no doubt that the wells are running dry. 'Someday we might look back with a curious nostalgia at the days when profligate homeowners wastefully sprayed their lawns with liquid gold to make the grass grow, just so they could then burn black gold to cut it down on the weekends,' says Michael E. Webber, associate director of the Center for International Energy and Environmental Policy at the University of Texas at Austin. 'Our children and grandchildren will wonder why we were so dumb.'

Water, water everywhere . . . or perhaps not

In 2008, an editorial in *Nature* outlined the problems humans have when it comes to water. More than a billion people in developing nations lack access to safe drinking water, and more than two billion have no proper sanitation. In the near future, water shortages are likely to spread into other key sectors, notably agriculture and energy.

This crisis is being driven partly by climate change, as rising temperatures lead to drier soils and less rainfall in many parts of the world. But there is the added pressure of increasing populations too, wrote *Nature*. 'As nations such as India and China grow more prosperous, for example, their citizens are switching to more protein-rich Western diets. It takes some 15,500 litres of water to produce a kilogram of industrial beef, ten times as much as is needed to produce 1 kilogram of wheat. These nations are likewise shifting their energy consumption towards intensities common in the developed world.'

The United States uses 500 billion litres of fresh water per day, around 40 per cent of its freshwater use, for cooling electric power plants. The country uses the same again for irrigation. 'The resulting pressures on water supplies are unrelenting. Global energy demand is projected to increase 57%

by 2030, and water demand for food production might easily double,' says *Nature*. 'By 2050, feeding the world's growing population may require some 3,000 cubic miles of water – the volume of Lake Superior – every year. Yet many of the world's rivers and lakes are already dramatically overused: China's Yellow River doesn't always reach the ocean, and Lake Mead in the American southwest could be dry by 2021 if water usage is not curtailed.'

In 2005, the United Nations reported that by 2050, more than 4 billion people would be living in nations defined as water-scarce or water-stressed, up from half a billion in 1995. A year later, the International Water Management Institute conducted a study showing that a third of the world's population was already living in water-scarce areas. When the same team had carried out the same analysis six years earlier, they had predicted that such a situation would not occur until 2025.

The IPCC's verdict on the effects of climate change on global water supply is also stark. 'Globally, the negative impacts of future climate change on freshwater systems are expected to outweigh the benefits,' it said. 'By the 2050s, the area of land subject to increasing water stress due to climate change is projected to be more than double that with decreasing water stress. Areas in which runoff is projected to decline face a clear reduction in the value of the services provided by water resources. Increased annual runoff in some areas is projected to lead to increased total water supply. However, in many regions, this benefit is likely to be counterbalanced by the negative effects of increased precipitation variability and seasonal runoff shifts in water supply, water quality and flood risks.'

Climate models for the rest of this century show that, because of global temperature increases due to greenhouse gas emissions, precipitation will increase in high latitudes and parts of the tropics, but will decrease in subtropical and lower mid-latitude regions, including the Mediterranean basin, the western USA, southern Africa and north-eastern Brazil. Extreme droughts will become ever more common, contributing to the disappearance of vast swathes of vegetation – in some scenarios, extreme drought will increase from one per cent of present-day land area to 30 per cent by 2100.

Which areas are under stress?

By 2025, nine countries in the eastern and southern parts of Africa will have only 1,000 cubic metres of water available per person per year. Twelve further countries on the continent will be limited to 1,000–1,700 cubic metres per year, and the population at risk of water stress could rise to 460 million, mainly in western Africa. 'In addition, one estimate shows the proportion of the African population at risk of water stress and scarcity increasing from 47% in 2000 to 65% in 2025,' said the IPCC in its 2008 report on climate change and water. 'This could generate conflicts over water, particularly in arid and semiarid regions.'

Around the Mekong Delta in Vietnam, a doubling of CO_2 in the atmosphere will cause enough climate change to trigger severe flooding in the wet season and the risk of extreme drought in the dry season.

Water availability in India is projected to decline from about 1,820 cubic metres per year per person in 2001 to 1,140 cubic metres per year in 2050, as a result of population growth alone.

133

However, a more pessimistic scenario has India reaching a state of water stress before 2025, when the availability is projected to fall below 1,000 cubic metres per person, according to the IPCC, due to climatic and demographic factors.

In South America, the number of people living in water-stressed environments (in other words, with supplies of less than 1,000 cubic metres per person per year) in the absence of climate change was estimated at 22.2 million in 1995. This number is expected to increase to between 12 and 81 million in the 2020s and to between 79 and 178 million in the 2050s.

The decline is visible everywhere. In 1940, the Chacaltaya glacier in Bolivia measured 0.22 km^2; by 2005, this had reduced to less than 0.01 km^2. Over the period 1992 to 2005, the glacier suffered a loss of 90 per cent of its surface area, and 97 per cent of its volume of ice.

As well as the problems described above, lack of fresh water is also a risk factor for bad health. 'Childhood mortality and morbidity due to diarrhoea in low-income countries, especially in sub-Saharan Africa, remains high despite improvements in care and the use of oral rehydration therapy,' says the IPCC. 'Climate change is expected to increase water scarcity, but it is difficult to assess what this means at the household level for the availability of water, and therefore for health and hygiene.'

What can we do?

Quite a lot, as it happens – the potential for savings, without hurting human health or economic productivity, is vast. 'Improvements in water-use efficiency are possible in every sector. More food can be grown with less water (and less water contamination) by shifting from conventional flood irrigation to drip and precision sprinklers, along with more accurately monitoring and managing soil moisture,' says Peter H. Gleick, president of the Pacific Institute, writing in *Scientific American*. 'Conventional power plants can change from water cooling to dry cooling, and more energy can be generated by sources that use extremely little water, such as photovoltaics and wind. Domestically, millions of people can replace water-inefficient appliances with efficient ones, notably washing machines, toilets and showerheads.'

According to *Nature*, the key to tackling the crisis in the most food-insecure parts of the world is to manage so-called 'green water', the abundant moisture that infiltrates the soil from rainfall, and that can be taken up by the roots of plants. 'Experts estimate that in regions such as sub-Saharan Africa, where more than 95% of crops are rain-fed, only 10–30% of the available rainfall is being used in a productive way. The fixes they suggest are decidedly low-tech: harvesting rainwater, planting roots deeper, better terracing, and switching from ploughing to tilling. Yet the potential gains could be enormous. In heavily irrigated regions such as south Asia, meanwhile, equally simple improvements in water usage could take the pressure off precious blue-water supplies, and hence drinking water.'

The biggest change, though, must be in the way we view water. No longer should it be seen as a free, limitless resource that falls from the skies and fills seas and rivers ready for our exploitation. Perhaps then we might use it more thoughtfully, so that in future there will be more to go around.

RESOURCE DEPLETION

The raw materials that keep 21st century life moving and communicating are largely invisible. Forget the stone, iron and fire of previous generations, today we depend on a set of materials from the most unfamiliar reaches of the periodic table of elements. Without them, we would not have electronics or drugs or plastics. And they are running out.

You know that your computer is made of plastic and aluminium or steel, but deep inside, there are minute traces of more curious elements. These elements make magnets stronger, so that your hard disk can be smaller; they make lasers more powerful and reliable so that they can be used everywhere with ease; and they speed up industrial chemical reactions so that the products we need are cheaper and easily available. Without these elements, modern society would not exist.

How bad is the problem?

Ever since the Industrial Revolution, the demand for raw materials has been shooting up. John E. Tilton, a mineral economist at the Colorado School of Mines, reckons that humans have used up more aluminium, copper, iron, steel, phosphate rock, sulphur, coal, oil, natural gas, and even sand and gravel over the past century than during all earlier centuries put together. Without this stuff, he says, modern life would be hard to imagine. And the pace of consumption is only getting faster.

Jared Diamond, a professor of geography at the University of California, Los Angeles, and author of *Collapse* and *Guns, Germs, and Steel*, estimates that the average rate at which people consume resources such as oil and metals, and produce waste including plastics and greenhouse gases, is about 32 times higher in North America, western Europe, Japan and Australia than it is in the developing world.

The world's fastest-growing economy, China, which has a population of 1.3 billion people, has started to consume at Western levels; if this continues, the world will run out sooner rather than later. 'Per capita consumption rates in China are still about 11 times below ours, but let's suppose they rise to [US] levels,' wrote Diamond in *The New York Times*. 'Let's also make things easy by imagining that nothing else happens to increase world consumption – that is, no other country increases its consumption, all national populations (including China's) remain unchanged and immigration ceases. China's catching up alone would roughly double world consumption rates. Oil consumption would increase by 106 per cent, for instance, and world metal consumption by 94 per cent.'

If the calculations included India as well as China, global consumption rates would triple. 'If the whole developing world were suddenly to catch up, world rates would increase elevenfold,' says Diamond. 'It would be as if the world population ballooned to 72 billion people.'

The question of resource depletion is not concerned with the things you might normally hear about – oil, for example – but rather with the stuff that underpins our hi-tech world. The idea of 'materials security' has become a prominent issue in the first decade of the 21st century.

Take helium, the second most abundant element in the universe. Its two stable isotopes, helium-3 and helium-4, are essential components in cryogenic technology. If you need a superconducting magnet, its temperature will have to be dropped to within a few degrees of absolute zero (–273.15°C), and for this you require supercooled helium. But even as use of this element is on the rise, the price has been kept artificially low by the US government, which controls the US National Helium Reserve near Amarillo, Texas. According to Nobel laureate Robert Richardson, who discovered some of liquid helium's more unusual properties, 'the world would run out in 25 years, plus or minus five years'.

An analysis by the Öko-Institut, funded by the United Nations and the European Union, highlighted several metals that will be needed for sustainable technologies of the future. These include tantalum, indium, ruthenium, gallium, germanium, cobalt, lithium, platinum and palladium for 'green' electronics, solar cells and batteries.

Exact information about how much of these critical elements there is in the world is often difficult to establish, subject to the secrecy of mining companies and metal traders. But it is clear that they are not infinite, and it is clear too that they will run out as the world's population develops.

And then there's the rare earth elements . . .

'Throughout history, political fights and international wars have often been waged over securing valuable resources such as oil, water and food,' declared a December 2010 editorial in the journal *Nature Photonics*. 'Now, a group of 17 elements in the periodic table known collectively as the "rare earths" (named so because they were first found within rare minerals buried deep underground) are at the centre of a political storm that threatens the photonics industry.'

The importance of these elements might not be immediately obvious, but our lives would not be the same without them, continued the editors of the journal. 'Elements such as erbium, ytterbium, yttrium, neodymium, thulium and europium are vital optically active ingredients at the heart of many lasers, optical amplifiers and phosphors. Put simply, rare earths transform otherwise benign crystals, glass fibres and thin films into materials that are capable of emitting and amplifying light.'

In other words, rare earth elements are essential components of the lasers that we now see everywhere, from supermarket checkouts to spacecraft. Beyond that, they are also used in magnets, batteries and lightweight metal alloys. Their specific chemical and physical properties make them useful in improving the performance of computer hard drives, catalytic converters, mobile phones, hi-tech televisions and sunglasses.

As technology advances, so the demand for the metals rises; in the past decade, their use has doubled. There are several kilograms of such elements in typical hybrid petrol-electric cars made by Toyota and Honda, a market that will expand in coming years.

Despite their name, rare earth elements are not actually all that rare. In a report published in 2010, the British Geological Survey put their natural abundance on the same level as copper or lead. The problem is access and supply: China has a near-monopoly on mining these elements. It owns 37 per cent of the world's estimated reserves, about 36 million tonnes, but controls more than 97 per cent of production. The former Soviet bloc has around 19 million tonnes and the US 13 million, with other large deposits held by Australia, India, Brazil and Malaysia.

The US House of Representatives is so worried about security of supply that it is considering legislation to try to end America's dependence on Chinese imports. The Mountain Pass mine in California, shut down in 2002 because of environmental and cost issues, is now to be reopened, along with potential mines at Bear Lodge in Wyoming and Bokan Mountain in Alaska.

Other sources, untapped as yet, include Greenland, which estimates suggest could meet 25 per cent of global demand for rare earth elements. South Africa also has potential for rich deposits, as do Malawi, Madagascar and Kenya.

What could happen?

Jared Diamond warns that the imbalance in general consumption between the developed and developing worlds will lead to trouble. 'People in the third world are aware of this difference in per capita consumption, although most of them couldn't specify that it's by a factor of 32,' he wrote in *The New York*

Times. 'When they believe their chances of catching up to be hopeless, they sometimes get frustrated and angry, and some become terrorists, or tolerate or support terrorists. Since September 11, 2001, it has become clear that the oceans that once protected the United States no longer do so. There will be more terrorist attacks against us and Europe, and perhaps against Japan and Australia, as long as that factorial difference of 32 in consumption rates persists.

'People who consume little want to enjoy the high-consumption lifestyle. Governments of developing countries make an increase in living standards a primary goal of national policy. And tens of millions of people in the developing world seek the first-world lifestyle on their own, by emigrating, especially to the United States and Western Europe, Japan and Australia. Each such transfer of a person to a high-consumption country raises world consumption rates, even though most immigrants don't succeed immediately in multiplying their consumption by 32.'

What can we do?

No atoms are destroyed when they are incorporated into appliances or used in industry (though helium is leached from the atmosphere into space). In theory, even the rarest elements can be recycled, but sometimes they can become so dispersed as to render them almost unobtainable.

Worldwide, more than 400 million tonnes of metal is recycled each year. Japan's National Institute for Materials Science believes its scientists can go further in extracting value from the country's discarded hi-tech gadgets: it has estimated that there is around 6,800 tonnes of gold (16 per cent of the world's reserves), 60,000 tonnes of silver (22 per cent) and 1,700 tonnes of indium (15.5 per cent) in Japanese 'urban mines'.

Replacement is also important. In the past two decades, flat-panel displays have overtaken the traditional cathode-ray screens in our living rooms. These screens are also used in mobile phones and other devices. But there is an impending bottleneck: each screen needs indium, an element that is notoriously hard to extract.

In 2002, Hideo Hosono and colleagues at the Tokyo Institute of Technology showed that they could use alumina and lime instead of indium, replacing a rare substance with the more abundant aluminium, calcium and oxygen. Similarly, catalysts such as platinum, palladium, rhodium, iridium and ruthenium are being replaced by iron, copper, zinc and manganese.

'Carbon is present abundantly and universally, and can be transformed into many kinds of compounds,' say Eiichi Nakamura and Kentaro Sato, both chemists at the University of Tokyo. 'It is therefore a very promising candidate as an ingredient in alternative materials for a range of functions. A pioneering example is the organic semiconductors that are used in OLEDs and thin-film solar cells. In that respect, the position of organic thin-film solar cells relative to silicon solar cells and compound solar cells should be assessed not merely in terms of our energy policy, but also in terms of the element strategy.'

Can we do it?

This is not the first time that humans have experienced a shortage of materials relying on certain elements, say Nakamura and Sato. 'About a century ago, when the advent of food shortages caused by the lack of nitrogen fertilizers became apparent, the advanced chemistry of that time averted disaster: the Haber–Bosch process saved the world from the crisis. The process enables ammonia and nitrate, which are essential for food production, to be synthesized from nitrogen in the air.'

Passing through the coming bottleneck will require a level of collective action that is nowhere yet in sight, believes Jeffrey D. Sachs, director of the Earth Institute at Columbia University and a special adviser to the former United Nations Secretary General Kofi Annan, writing in the journal *Science*. 'Budget funding for the future technologies that could underpin sustainable development is a small fraction of military spending, and only a slight part of that spending is directed at the health, energy, and environmental needs of the world's poorest people.'

Although the depletion of rare elements could trigger a new conflict, it could also serve as an opportunity to make a dream come true, to create new science and technology. Diamond is cautiously optimistic, citing rising awareness of environmental issues around the world in recent years as his reason for hope. 'The world has serious consumption problems, but we can solve them if we choose to do so.'

ENVIRONMENTAL COLLAPSE

Humans need food, drink, energy and raw materials to live. We need space to create cities, grow food and build factories. All of it comes, ultimately, from the Earth, and the success of our species is crowding out hundreds of thousands of others.

We cut down swathes of forest, we fish the oceans empty of life, we poison and concrete over the land. We drive tens of thousands of animal and plant species to extinction as we outcompete them for basic resources such as energy or, simply, space. We are living through a mass extinction of species that has not been matched since the age of the dinosaurs, but this time it is being caused by humans.

This loss of biodiversity is not just a concern for tree-huggers and bird-lovers. A decline in the world's species can cause ecosystems to become stressed, degraded and liable to collapse. 'This threatens the continued provision of ecosystem services, which in turn further threatens biodiversity and eco-system health,' said the environmental charity WWF (World Wide Fund for Nature) in its Living Planet report of 2010, a stocktake of the total human footprint on the world. 'Crucially, the dependency of human society on ecosystem services makes the loss of these services a serious threat to the future well-being and development of all people, all around the world.'

Billions of years of evolution on Earth have created a web of life forms that are dependent on each other for survival. The animals and plants of the world are the foundation of complex ecosystems, and we humans are more critically tied into and dependent on nature than most of us care to acknowledge. If it fails, so do we.

What are we doing to the world?

Humans degrade our environment in multiple ways. We destroy and cut up natural habitats: forests might be cleared to make way for agricultural land or a new town or industrial plant; a river might be dammed to build a hydroelectric power plant or to improve the irrigation of nearby fields. We over-exploit animals and plants in the wild for food, raw materials, medicines or sport. We move species from one habitat to another, causing major problems for local varieties as they are forced to compete for resources or cannot fight off diseases brought in by the interloper. And through pollution and climate change, we are poisoning and changing the environment of the whole world.

In the Millennium Ecosystem Assessment of 2005, researchers concluded that changes in biodiversity in the past 50 years had been more rapid than at any other time in human history, and that the drivers of these changes were showing no evidence of decline.

'We're living in a time of mass extinctions that exceeds the fossil record by a factor of 10,000,' said Stephen Petranek, former editor of *Discover* magazine, in a lecture for TED.com.

'We have lost 25% of the unique species in Hawaii in the last twenty years, California is expected to lose 255 of its species in the next forty years. Somewhere in the Amazon forest is the marginal tree. You cut down that tree, the rainforest collapses as an ecosystem. There's really a tree like that out there. That's really what it comes to. And when that ecosystem collapses, it could take a major ecosystem with it, like our atmosphere.'

In 2010, WWF reported on the health of the world's bio-diversity. Populations of species in the tropics, it found, were falling through the floor, yet human demand for natural resources was shooting up. We are using a planet and a half's worth of resources, said its Living Planet report.

In the past forty years, our consumption of nature has doubled, while the 'Living Planet index' – a measure of the decline and increase of almost 8,000 populations of more than 2,500 species of marine, fluvial and land species – has dropped by 60 per cent in the tropics and 30 per cent overall.

'There is an alarming rate of biodiversity loss in low-income, often tropical countries while the developed world is living in a false paradise, fuelled by excessive consumption and high carbon emissions,' said Jim Leape, director general of WWF International, when the 2010 report was launched.

The biggest declines in biodiversity were found in the lowest-income countries, with an almost 60 per cent overall drop in the past forty years.

The WWF blamed unsustainable consumption in wealthier nations, which depleted the natural resources of poorer countries. The report found that the ten countries with the largest ecological footprints per person were the United Arab Emirates, Qatar, Denmark, Belgium, the United States, Estonia, Canada, Australia, Kuwait and Ireland. Furthermore, the thirty-one OECD (Organization for Economic Co-operation and Development) countries, which include the world's richest economies, account for nearly 40 per cent of the global footprint. Twice as many people live in the emerging economies of Brazil, Russia, India and China, and their ecological footprint per person could overtake that of the OECD countries if they were to follow the same path of development.

If we continue living beyond the Earth's limits, by 2030 we will need the resources of two Earths to keep up with annual demand. 'The report shows that continuing the current consumption trends would lead us to the point of no return. 4.5 Earths would be required to support a global population living like an average resident of the US,' said Leape.

What is disappearing?

As of 2010, the golden-headed langur, which is found only on the island of Cat Ba in north-eastern Vietnam, was down to sixty to seventy individuals. There were fewer than a hundred northern sportive lemurs left in Madagascar, and around 110 eastern black-crested gibbons in north-eastern Vietnam. The Sumatran orang-utan is down to around 6,600 due to fragmentation of their habitats and the removal of forest to make way for agricultural uses such as palm oil plantations.

The World Conservation Union (IUCN) reckons that almost half the world's primate species – which include apes, monkeys and lemurs – are threatened with extinction due to the destruction of tropical forests and illegal hunting and trade. The plight of primates from Madagascar, Africa and Asia to Central and South America is desperate – 48 per cent of the world's 634 primate species are threatened, with many at imminent risk of extinction. And when a population is small, the disasters are always big – a tropical cyclone could easily wipe out the last few hundred individuals.

In the seas, sharks are disappearing fast. The scalloped hammerhead shark has declined by 99 per cent over the past thirty years in some parts of the world, and has been declared globally endangered on the IUCN's Red List of species at risk, which contains more than 130 varieties of shark. Populations in the north-west Atlantic Ocean have declined by an average of 50 per cent since the early 1970s. In 2007, twenty-one shark-fishing nations reported catching more than 10,000 tonnes of shark. The top five – Indonesia, India, Taiwan, Spain and Mexico – accounted for 42 per cent of this figure.

Sharks are especially vulnerable because they can take decades to mature and they produce few young. 'Sharks are definitely at the top of the list for marine fishes that could go extinct in our lifetimes,' says Julia Baum of the Scripps Institution of Oceanography in California and a member of the IUCN shark specialist group. 'If we carry on the way that we are, we're looking at a really high risk of extinction for some of these shark species within the next few decades.'

There are countless other examples of animals and plants threatened in some way as humans spread: tigers, coral reefs, gorillas, northern white rhinoceroses, axolotls, leatherback sea turtles, Chinese alligators, Hawaiian crows and snow leopards are just a minuscule portion of the species under pressure.

But what has all this got to do with people?

Excessive fishing in recent decades has caused a 90 per cent decline in shark populations across the world's oceans, and up to 99 per cent along the US east coast. This has already started affecting the way people live. After a collapse of shark numbers in 2000, the sharks' prey, cownose rays, exploded on US shores. The rays in turn decimated the bay scallop populations around North Carolina in 2004, disrupting the fisheries there and shutting down a local economy that had lasted for well over a century.

In Costa Rica, it is well known that coffee farms located within close distance of forests produce better coffee and can increase yields by 20 per cent. The pollination service from forest areas translated into an income of $60,000 per year for one farm.

In Ecuador, 80 per cent of the water for the capital, Quito, comes from three protected areas that are threatened by human activities, according to WWF, such as logging and conversion into farms.

According to the World Health Organization, natural compounds from animals, plants and microorganisms are an important source of drugs to treat human diseases. Half of all current medical compounds start life as natural products such as aspirin, digitalis and quinine.

Can we reverse the decline?

The WWF warns that current rates of consumption and degradation of the natural environment will lead to ecosystem collapse within fifty years. 'We must balance our consumption with the natural world's capacity to regenerate and absorb our wastes. If we do not, we risk irreversible damage,' says Leape.

There are international schemes afoot to try and stem some of the losses. The UN's REDD programme (Reducing

Emissions from Deforestation and Forest Degradation in Developing Countries), which is proposed as part of any global deal to tackle climate change, will be crucial in maintaining falling primate populations. The idea is that rich countries would pay developing countries to maintain their forests, therefore locking in the carbon and preventing further greenhouse gas emissions. In an enhanced version of the idea, developing countries will be incentivized to plant more trees, expanding their forest areas.

In 2010, environment ministers from almost 200 countries responded to the catastrophic situation by agreeing plans to try and stop the worst loss of life on Earth since the demise of the dinosaurs. At a meeting in Nagoya, Japan, they resolved to halve the destruction of natural habitats and expand nature reserves to 17 per cent of the world's land area by 2020, up from less than ten per cent today. These Aichi targets, named after the region around Nagoya, mean more refuges for sealife too, with an increase of marine protected zones to cover ten per cent of the world's oceans, up from today's one per cent.

The Aichi targets will come into force in 2020, and require all signatories to draw up national biodiversity plans. In combination, it is hoped that these plans will, among other things, stop overfishing, reduce pollution and control invasive species.

But agreeing targets is one thing, implementation quite another. After the resolutions were passed in Nagoya, environmental journalist George Monbiot was sceptical about whether they would change anything. 'The draft saw the targets for 2020 that governments were asked to adopt as nothing more than "aspirations for achievement at the global level" and a "flexible framework", within which countries can do as they wish. No government, if the draft has been approved, is obliged to change its policies,' he wrote.

'It strikes me that governments are determined to protect not the marvels of our world but the world-eating system to which they are being sacrificed; not life, but the ephemeral junk with which it is being replaced. They fight viciously and at the highest level for the right to turn rainforests into pulp, or marine ecosystems into fishmeal. Then they send a middle-ranking civil servant to approve a meaningless and so far unwritten promise to protect the natural world.'

It suits governments to let us trash the planet, said Monbiot. 'It's not just that big business gains more than it loses from converting natural wealth into money. A continued expansion into the biosphere permits states to avoid addressing issues of distribution and social justice: the promise of perpetual growth dulls our anger about widening inequality. By trampling over nature we avoid treading on the toes of the powerful. A massive accounting exercise, whose results were presented at the meeting in Japan, has sought to change this calculation.'

Ecosystems are still under pressure, and it is a problem we are far from starting to solve.

RISING SEA LEVELS

Rising sea levels are one of the clearest indications that the Earth's climate is changing. Scientists might still be engaged in finding out how fast, how much and where the effects will be most keenly felt, but there is little doubt that it is already having an impact on our seas.

Low-lying areas of the world are being slowly subsumed by water, storms are on the rise, and coastal towns and cities are experiencing bigger floods more often. The worst part is, anything that has happened so far is just the beginning. Climate change has much more in store for those who live by the sea and, perhaps later this century, for people living further inland too.

We know that sea levels will rise over the next century as the world continues to warm, and we also know that this means death, devastation and the end of livelihoods for millions of people.

The Earth has easily enough water to drown most of its human population. Worse still, we have built most of our biggest and most important cities near oceans, rivers and seas. A major increase in sea level would change every one of our lives. The higher the seas rise, the less food we can grow, the fiercer storms will be and the smaller the human population will become. The question is, how far will we go in forcing the Earth to pull out all the stops?

Why does the sea-level change?

Water exists in three distinct forms on Earth: the familiar liquid that fills the seas and washes on to coastlines; the vapour in the air; and the vast sheets of ice covering continents and floating on oceans. The balance of these three in any part of the world depends in the short-term on local temperature and weather. Long-term, it depends on climate and the flow of energy around the planet.

The volume of water in the sea can go up for two reasons: water expands as it gets warmer; or extra water enters the sea because ice sheets melt. Both these things have been happening for the past century, partly due to natural cycles, but also driven by man-made global warming, as we burn fossil fuels and throw greenhouse gases into the atmosphere.

'Measurements from tide gauge stations around the world show that the global sea level has risen by almost 20 cm since 1880,' says Stefan Rahmstorf, a climate scientist and ocean-ographer at the Potsdam Institute for Climate Impact Research. 'Since 1993, global sea level has been measured accurately from satellites; since 1993 figures have shown levels rising at a rate of 3.2 cm per decade.'

One of the biggest ice sheets in the world sits near the Arctic Circle, on the land mass of Greenland. The land here was not always covered – around 60 million years ago, the Earth was far warmer than today, and Greenland was a grassy tundra populated by ancient mammals. Today's ice sheet, which is three kilometres (almost two miles) thick in some parts, was formed in the last ice age, around 20,000 years ago, and the reason it is still there is because average temperatures between then and now have not risen high enough to melt it. Any ice that has melted into the surrounding sea has been replaced by regular snowfall in that time.

But Greenland's equilibrium might now be on the rocks. Under the worst climate scenarios predicted for this century, the average temperature around Greenland might increase by 8°C by 2100. If all the ice there melted into the sea, global sea levels could rise by around seven metres.

Even if the average temperature in Greenland increased by 3°C, its ice sheet would largely disappear, though it would take a long time – around 1,000 years – for it to melt completely. In 2004, Jonathan Gregory of the University of Reading showed that by 2100, concentrations of greenhouse gases would probably have reached levels sufficient to raise the temperature past this warming threshold.

At the other end of the world is an even bigger continent of ice – Antarctica. Overall it is more stable than Greenland, but if just a small part of the west Antarctic ice sheet were to melt (which is not an unreasonable possibility, given its various major collapses during the past decade), the world's sea levels would jump by up to five metres.

Of course, if temperatures keep going up, ever more ice will fall into the sea and the water will continue to rise. If all the ice in the world melted, from mountain ranges, Greenland and Antarctica, the result would be catastrophic. 'The land-based ice sheets of Greenland and Antarctica hold enough water to raise global sea level by more than 200 feet,' says Robin Bell, an expert on Antarctica at Columbia University's

Lamont-Doherty Earth Observatory. A sea-level rise of this order would decimate the world's big cities and much of their populations along with it.

Which areas are at risk from higher water?

Any low-lying countries and anyone living by the coast is at risk from a global rise in sea levels. Major cities including London and New Orleans already need protection against storm surges, and would need even better defences if sea levels rose. A one-metre rise would swamp cities on the eastern seaboard of the US. A six-metre rise would drown most of Florida.

Places such as the Maldives, Tuvalu (where the highest point is currently four metres above sea level) and scores of Pacific islands would be subsumed entirely with sea-level rises of not much more than a few metres.

In his economic analysis of the impacts of climate change, Nicholas Stern of the London School of Economics calculated that 200 million people live within one metre above the present sea level. This includes eight out of ten of the world's largest cities, and all the megacities of the developing world.

'Miami tops the list of most endangered cities in the world, as measured by the value of property that would be threatened by a three-foot rise,' say Rob Young, a geoscientist at Western Carolina University, and Orrin Pilkey of Duke University. 'This would flood all of Miami Beach and leave downtown Miami sitting as an island of water, disconnected from the rest of Florida.' Other threatened US cities include New York, Newark, New Orleans, Boston, Washington, Philadelphia and San Francisco. Outside North America, major risks of flood and water rise are faced by Osaka, Kobe, Tokyo, Rotterdam, Amsterdam and Nagoya.

Vivien Gornitz of Columbia University's Center for Climate Systems Research worked out that rising oceans would eat away almost 2,400 km (1,500 miles) of shoreline in and around New York, which is home to around 20 million people. 'Sea level has already climbed around 27 cm in New York City and 38.5 cm along the New Jersey coast during the 20th century,' she says. 'These local rates exceed the global average of 10–25 cm per century because the East Coast is slowly sinking, as the earth's crust continues to readjust to the removal of the ice from the last glaciation, around 15,000 years ago. But present rates of sea level rise could accelerate several-fold, as mountain and polar glaciers melt and upper ocean layers heat up and expand, due to global warming.'

Rising seas also increase the chances of storm surges and dangerous floods. The severity of a flood is often defined in terms of the probability of its occurring once every hundred years. At the beginning of the 21st century, the height of a hundred-year flood around New York was just under 3 metres (10 feet), an event that would devastate the city and its surrounding area. According to Gornitz, if the seas rose, smaller surges could produce hundred-year floods. By the 2080s, given present rates of sea-level rise, the likelihood of such an event would be once in fifty years. It would rise to once every four years in the very worst-case scenario.

In eastern Asia, 18 million people live in the Vietnamese portion of the Mekong Delta, which accounts for around a fifth of Vietnam's population and more than 40 per cent of its cultivated land surface. They grow half the country's rice, produce 60 per cent of the fish and seafood and harvest 80 per cent of the fruit crop. A one-metre sea-level rise would displace 7 million people, and a two-metre rise would displace twice that number. Not to mention the devastating impact of increased floods, beyond anything the locals are used to. The government, mindful of the coming changes, is already moving people away from some parts of the Mekong River's main branch in An Giang province.

Is it likely?

Back in 2007, the Intergovernmental Panel on Climate Change (IPCC) issued its fourth report on the impacts of global warming, in which it suggested that by 2100, seas might rise by between 180 and 590 mm. Subsequent research, which took into account more advanced models of how ice sheets move and melt, suggested that the end-of-century rise might well be twice the IPCC's estimate, with a likely limit of around two metres.

In the 20th century, sea-level rise was primarily due to the expansion of ocean water as it warmed up. Contributions from melting mountain glaciers and the large ice sheets were minor components. 'But most climate scientists now believe that the main drivers of sea level rise in the 21st century will be the melting of the West Antarctic Ice Sheet (a potential of a 16-foot rise if the entire sheet melts) and the Greenland Ice Sheet (a potential rise of 20 feet if the entire ice cap melts),' say Young and Pilkey. 'The nature of the melting is non-linear and is difficult to predict.'

There is an added complication, revealed by scientists in recent years. Underneath ice sheets there are complex systems of rivers, lakes and meltwater, all of which can enable the flow of vast chunks of ice towards the ocean. 'For millennia, the outgoing discharge of ice has been balanced by incoming snowfall,' says Robin Bell. 'But when warming air or surface meltwater further greases the flow or removes its natural impediments, huge quantities of ice lurch seaward. Models of potential sea-level rise from climate change have ignored the effects of subglacial water and the vast streams of ice on the [overall] flow of ice entering the sea.'

It is fair to say that the global sea level is rising much faster than expected. 'A commission of 20 international experts, called on by the Dutch government to help plan its coastal defences, has recently given a high-end estimate of 55 cm to 110 cm by 2100,' says Stefan Rahmstorf. 'Equally important, this commission has highlighted the fact that sea level rise will not stop in the year 2100. By 2200, they estimate a rise of 1.5 to 3.5 m unless we stop the warming. This would spell the end of many of our coastal cities.' And that's just the scenario the scientists can predict.

THE GULF STREAM SHUTS DOWN

Our planet's climate might seem slow-moving. But the records show that it has a disturbing ability to make abrupt shifts from one climate to another in record time. What happens to life if the world around it changes overnight?

Climate change is usually understood to mean an overall gradual rise in temperature around the world. Often the term is used interchangeably with global warming. In any case, its implications are clear: environmental change, pressure on resources and a gradual transition to a warmer world.

But this description masks some of the more horrifying potential effects. Evidence gathered from ancient episodes of climate change on Earth reveals that our planet has gone from warm to cold in a matter of just a few years, rather than the centuries scientists usually expect for climactic changes.

The disaster scenario that keeps scientists awake at night concerns a huge current of water that flows around our oceans, carrying energy from one part of the Earth to another. The thermohaline circulation is indescribably crucial in keeping temperature and energy balanced across the globe and making the northern latitudes, including Europe, more habitable. Unfortunately, it seems as though it would not take much to knock this conveyor belt of water out of kilter and ruin the lives of billions of people.

If the thermohaline circulation were to shut down, the heat-bearing Gulf Stream would stop, and winters in the North Atlantic, Europe and North America would become twice as severe as anything on record, with average temperatures dropping by up to 5°C, moisture in soil falling and winds becoming more intense. Since these are areas that provide a significant fraction of the world's food, our ability to support the global human population would plummet.

'Fossil evidence clearly demonstrates that Earth's climate can shift gears within a decade, establishing new and different patterns that can persist for decades to centuries,' says Robert B. Gagosian, president of the Consortium for Ocean Leadership and former director of the Woods Hole Oceanographic Institution in Massachusetts. 'In addition, these climate shifts do not necessarily have universal, global effects. They can generate a counterintuitive scenario: Even as the Earth as a whole continues to warm gradually, large regions may experience a precipitous and disruptive shift into colder climates.'

Geological records confirm that an abrupt thermohaline-induced climate change would generate severe winters in the North Atlantic. A few bad winters might be inconvenient, but we are able to deal with them. A persistent series of them over decades or even a century, however, could cover countries in ice, make rivers freeze, and cause sea ice to grow and spread. The Western world as we know it, the centre of so much commerce, agriculture and political power, would be uninhabitable.

What is the thermohaline circulation?

The Sun shines on to the Earth at different intensities at different latitudes. The equator gets more sunshine than the poles, and global ocean currents move the energy around the Earth. One result of the equatorial sunshine is that the warmer water here evaporates more, leaving behind a saltier sea. The thermohaline circulation then moves huge volumes of warm, salty water from the tropics up the east coast of the US and on to Europe.

'This oceanic heat pump is an important mechanism for reducing equator-to-pole temperature differences,' says Gagosian. 'It moderates Earth's climate, particularly in the North Atlantic region. Conveyor circulation increases the northward transport of warmer waters in the Gulf Stream by about 50 per cent. At colder northern latitudes, the ocean releases this heat to the atmosphere – especially in winter when the atmosphere is colder than the ocean and ocean–atmosphere temperature gradients increase. The conveyor warms North Atlantic regions by as much as 5°C and significantly tempers average winter temperatures.'

Around the waters of the northern Atlantic – the Labrador, Irminger and Greenland Seas – the thermohaline circulation helps to release large amounts of heat into the atmosphere. After it passes through these seas, the cold winds around Iceland cool the water, which sinks and moves south, eventually flowing around the Antarctic. As the water sinks in the North Atlantic, salty tropical surface waters are drawn northwards to replace it.

It is a truly huge movement of water and energy. But records contained in ice cores and sediments show that this conveyor belt has not run steadily all the time. 'Variations in the conditions governing the density of high-latitude surface waters can lead to abrupt reorganizations of the ocean's circulation. The surprise revealed to us by the climatic record is the extent, rapidity, and magnitude of these atmospheric changes,' wrote Wallace Broecker, a climatologist at the Lamont-Doherty Earth Observatory of Columbia University, in a 1997 paper in *Science*.

How could it be shut down?

The thermohaline circulation operates because of temperature and salt differences in the world's oceans, which act like natural pumps to move the water from one place to another. Disrupt any part of that, and the overall conveyor belt will be adversely affected.

'Salty water is denser than fresh water. Cold water is denser than warm water. When the warm, salty waters of the North Atlantic release heat to the atmosphere, they become colder and begin to sink,' says Gagosian. 'If cold, salty North Atlantic waters did not sink, a primary force driving global ocean circulation could slacken and cease. Existing currents could weaken or be redirected. The resulting reorganization of the ocean's circulation would reconfigure Earth's climate patterns.'

The way to mess up the system is to add large amounts of fresh water into the North Atlantic part. This would dilute the salinity of the water, and at a certain threshold, it would not be dense enough to sink. This part of the conveyor belt would therefore stop.

A report by Peter Schwartz and Doug Randall for the Climate Institute in Washington DC plays out a scenario of what would happen next. 'The North Atlantic Ocean continues to be affected by fresh water coming from melting glaciers, Greenland's ice sheet, and perhaps most importantly increased rainfall and runoff. Decades of high-latitude warming cause increased precipitation and bring additional fresh water to the salty, dense water in the North, which is normally affected mainly by warmer and saltier water from the Gulf Stream. That massive current of warm water no longer reaches far into the North Atlantic. The immediate climatic effect is cooler temperatures in Europe and throughout much of the Northern Hemisphere and a dramatic drop in rainfall in many key agricultural and populated areas. However, the effects of the collapse will be felt in fits and starts, as the traditional weather patterns re-emerge only to be disrupted again – for a full decade.'

It has happened before. When the Earth came out of its most recent ice age, around 13,000 years ago, the thermohaline circulation was disrupted. The resulting cold period, called the Younger Dryas, lasted more than 1,000 years, with icebergs as far south as Portugal.

Another rapid change in ocean circulation occured 8,200 years ago, leading to widespread drought in the American West, Africa and Asia. 'Regional cooling events also have been linked with changes in the Southwest Asian monsoon, whose rains are probably the most critical factor supporting civilizations from Africa to India to China,' says Gagosian.

The effects of a shutdown today

Schwartz and Randall's scenario shows that there would be global effects if the thermohaline circulation were to shut down. Annual average temperatures would drop across Asia, North America and northern Europe, and droughts would persist for at least a decade in important agricultural and population centres. Temperatures would rise, however, across Australia, South America and southern Africa.

Winter storms and winds would intensify across western Europe and the northern Pacific. Access to energy supplies would be disrupted because of extensive sea ice and storms. The combination of wind and drought would cause widespread dust storms and loss of soil. By the end of a decade, the majority of Europe would feel more like modern Siberia.

'As global and local carrying capacities are reduced, tensions could mount around the world, leading to two fundamental strategies: defensive and offensive,' wrote Schwartz and Randall. 'Nations with the resources to do so may build virtual fortresses around their countries, preserving resources for themselves. Less fortunate nations, especially those with ancient enmities with their neighbors, may initiate struggles for access to food, clean water, or energy. Unlikely alliances could be formed as defense priorities shift and the goal is resources for survival rather than religion, ideology, or national honor.'

In the US, the cold, windy weather would make it more difficult to grow food. In China, the normally reliable monsoon rains would regularly fail, leading to widespread famine.

'The changing weather patterns and ocean temperatures affect agriculture, fish and wildlife, water and energy. Crop yields, affected by temperature and water stress as well as length of growing season, fall by 10–25% and are less predictable as key regions shift from a warming to a cooling trend. As some agricultural pests die due to temperature changes, other species spread more readily due to the dryness and windiness – requiring alternative pesticides or treatment regiments. Commercial fishermen that typically have rights to fish in specific areas will be ill equipped for the massive migration of their prey,' continue Schwartz and Randall.

'With only five or six key grain-growing regions in the world (US, Australia, Argentina, Russia, China, and India), there is insufficient surplus in global food supplies to offset severe weather conditions in a few regions at the same time – let alone four or five. The world's economic interdependence makes the United States increasingly vulnerable to the economic disruption created by local weather shifts in key agricultural and high population areas around the world. Catastrophic shortages of water and energy supply – both of which are stressed around the globe today – cannot be quickly overcome.'

Is it likely?

'The fate of the thermohaline circulation will be decided by Greenland,' the climate scientist Stefan Rahmstorf told *Nature*. 'If that goes quickly it will be bad news for the deep-water formation. But if Greenland is stable, the risk of shutting down the circulation completely is very small.'

In a 2002 paper, scientists found that fresh water had been entering the North Atlantic for the past forty years, and at an increased rate in the past decade. Since the mid-1960s, they discovered, the seas feeding the North Atlantic had steadily become less salty.

Will this result in a slowdown or a stop to the conveyor belt? 'The short answer is: We do not know. Nor have scientists determined the relative contributions of a variety of sources that may be adding fresh water to the North Atlantic. Among the suspects are melting glaciers or Arctic sea ice, or increased precipitation falling directly into the ocean or entering via the great rivers that discharge into the Arctic Ocean. Global warming may be an exacerbating factor,' says Gagosian.

Abrupt regional cooling, he adds, and gradual global warming can unfold simultaneously. 'Indeed, greenhouse warming is a destabilizing factor that makes abrupt climate change more probable. A 2002 report by the US National Academy of Sciences said, "available evidence suggests that abrupt climate changes are not only possible but likely in the future, potentially with large impacts on ecosystems and societies".'

SNOWBALL EARTH

Our world was once covered in ice. This was not just any old ice age, but something far more severe: glaciers stretched down from the poles to the equator, and the entire surface of the ocean was solid. Our planet, 700 million years ago, was a cosmic snowball hurtling through space.

Overhead, the sky bore only the faintest wisps of cloud – frozen carbon dioxide crystals rather than water, which had stopped circulating around the world due to the freezing temperatures.

For the life forms that had been developing at this time, 'Snowball Earth' was a disaster. Until then, our planet had been just warm enough, with exactly the right chemical conditions. Multicellular life forms were beginning to emerge, and mitochondria, tiny powerhouses that sit inside our body cells and turn our food into usable energy, had started to develop symbiotic relationships with the simple organisms that would one day evolve into the animals and plants we see today.

In a flash, all of it was gone. Nothing living on or near the surface survived. If the −50°C temperatures did not stop life flourishing, the lack of liquid water would have been a problem. Only the most primitive of life forms – some algae and bacteria – had any chance under the kilometre-thick ice covering, congregating around the hydrothermal vents and the relatively warm water around undersea volcanoes.

Snowball Earth devastated life millions of years ago; if a similar scenario appeared today, it would have far worse effects. Imagine our cities trapped under glaciers, all our infrastructure destroyed or useless, and access to water severely limited. Ice would cover the fields we use to grow crops, our sources of energy would be compromised, and billions of people would die of cold and hunger. For the Earth, it would be just a blip in its climate before some future return to warmer conditions. But that blip would spell the end of our civilization.

The Snowball Earth hypothesis

There have been many ice ages in the Earth's past, when the overall temperature dropped and large parts of the world were covered in ice. But in the past few decades, scientists have un-covered evidence of a far more extreme version of the freeze, one that puts all other known ice ages into the shade.

The first of these extreme glaciations is thought to have happened around 2.2 billion years ago, when the planet was half its current age. Another series of extreme freezes began around 700 million years ago and ended more than 130 million years later, at the end of the Proterozoic Eon.

At the time of the last Snowball Earth, our planet looked very different from the way it does today. The supercontinent Rodinia had just broken up into a set of land masses that were clustered around the Equator, and the Sun was giving out around 6 per cent less light.

The ice started to grow because the Earth's natural greenhouse effect failed. Because of their molecular structure, greenhouse gases in the atmosphere absorb some wavelengths of high-energy electromagnetic radiation coming from the Sun, but allow others (mainly visible light) to pass through. They also trap some of the heat that is radiated towards space from the surface of the Earth, making the surface warmer than it would be if there was no atmosphere.

The key greenhouse gas is carbon dioxide. Today we worry about rising levels of CO_2, and its potential to cause a gradual warming of the Earth's climate. Scientists think that the drop in global temperatures before the last Snowball Earth was partly due to a large-scale removal of CO_2 from the atmosphere.

They explain it like this: areas that had been previously landlocked in Rodinia were now closer to oceans and moisture, thanks to the break-up of the supercontinent. This created more rainfall, which pushed up a natural process called silicate weathering, where rainwater absorbs CO_2 from the air to create carbonic acid, which then disintegrates rocks over geological timescales to create soil. As rain increased, more CO_2 was absorbed from the air and more rock was slowly dissolved.

Several million years of excessive rain meant that CO_2 levels became low enough to compromise the greenhouse effect. The Earth began to freeze. Ice sheets started to stretch out from the poles, and the runaway drop in temperature was exacerbated by the bright white surfaces of the glaciers, which reflected away more of the Sun's incoming rays than the darker land or water underneath. With less greenhouse gas in the atmosphere to trap that reflected radiation, the energy simply disappeared back into space and was unavailable to keep the Earth warm.

When ice formed at latitudes lower than around 30° north, the fraction of light being reflected back into space rose at an ever faster rate because direct sunlight was striking a larger surface area of ice per degree of latitude. 'The feedback became so strong in this simulation that surface temperatures plummeted and the entire planet froze over,' say Paul Hoffman and Daniel Schrag, geologists at Harvard University and experts in the evolution and impact of Snowball Earth. For the next 30 million years, the perfect storm of bright white glaciers and lack of greenhouse effect continued to keep temperatures on the surface down to −50°C at the poles and −30°C at the tropics.

After the ice

The extreme conditions of Snowball Earth were described in 1992 by Joe Kirschvink, a specialist in palaeomagnetism at the California Institute of Technology, though scientists did not rush to support him at the time. It seemed unthinkable to them that the Earth's climate would allow glaciers to develop at the tropics, and if the process was self-reinforcing, how would it ever end? Why is the Earth not still a snowball today?

Kirschvink answered the second question by pointing out that the Earth's shifting tectonic plates would continue to build volcanoes and to pump CO_2 into the air. At the same time, since the world was covered in ice, the normal chemical cycles that removed the CO_2 from the atmosphere (namely, rain) would have stopped.

All of this would mean that the CO_2 would start to accumulate during the millions of years of freeze, and at the point when the levels of the gas had increased 1,000-fold, the greenhouse effect would return to melt the ice. 'The thaw takes only a few hundred years, but a new problem arises in the mean-time: a brutal greenhouse effect,' say Hoffman and Schrag. 'Any creatures that survived the icehouse must now endure a hothouse. As improbable as it may sound, we see clear evidence that this striking climate reversal – the most extreme imagin-able on this planet – happened as many as four times between 750 million and 580 million years ago.'

As the tropical oceans thawed after the last Snowball Earth, the water would have evaporated and, together with CO_2, made the Earth's greenhouse effect even more intense. Temperatures at the equator would have reached 50°C and rainfall would have been incessant. The hot seas would have supported powerful hurricanes.

After millions of years in deep freeze, the Earth would have woken up, and as it got back to a normal climate, conditions became perfect for a boom in life, culminating in the so-called Cambrian explosion between 575 and 525 million years ago, when there was a huge upsurge in the number of life forms on Earth.

How did it affect life?

The snowball itself killed almost everything. But the evidence in the geological record shows that once the last extreme freeze had thawed out, the first multicellular animals appeared. How did such an extreme event kick-start the evolution of life on Earth?

Geologists think that the explosion happened because once the snowball was gone, any organism that had clung on to life for the millions of years of the freeze was suddenly presented with vast empty environments in which to expand. As the glaciers advanced during the freeze, they carried the land's topmost layer of minerals and nutrients (from the soil and rocks) with them. Millions of years later, as the snowball melted and the glaciers retreated, these minerals and nutrients were released back into the oceans, feeding ever-larger numbers of oxygen-producing algae, which in turn created the atmospheric conditions needed for the evolution of animals.

Extreme freezing might well have killed almost all life on Earth, but it also created the conditions for the diversity of living things we see around us today.

Could it happen again?

Today, the world's continents are spread more evenly around the planet, so there is less chance of the intense evaporation and rainfall needed to clean the air of CO_2. In addition, our Sun is shining far brighter and hotter than it was at the start of the last Snowball Earth.

But Hoffman and Schrag say that we should be wary of the planet's capacity for extreme change. 'For the past million years, the earth has been in its coldest state since animals first appeared, but even the greatest advance of glaciers 20,000 years ago was far from the critical threshold needed to plunge the earth into a snowball state.'

We are more than 80,000 years from the peak of the next ice age, but the geologists say it is difficult to predict where the Earth's climate will drift over millions of years. 'If the trend of the past million years continues and if the polar continental safety switch were to fail, we may once again experience a global ice catastrophe that would inevitably jolt life in some new direction.'

CHEMICAL POLLUTION

In some ways, it sounds like the start of an apocalypse already: female pseudo-hermaphrodite polar bears with penis-like stumps; panthers with atrophied testicles; male trout and roach with eggs growing in their testes. This is not the line-up for some macabre circus, but the end of a long chain of events that starts with the materials that build modern society.

The way we grow our food, make the clothes we wear, build and maintain our cities – all of it is poisoning our planet.

Pesticides, radioactive waste, heavy metals, exhaust gases and excess fertilizers are just a few of the copious toxic chemicals that we dump into our rivers, oceans and air. The sheer size of the Earth has allowed us to turn a blind eye to the effects of all this pollution – clouds of gas dissipate into the atmosphere, effluent and industrial sludge blends into rivers, and pesticides leach away into the ground after they have done their job.

Luckily for us (and the planet), most of the things we dump will become diluted beyond danger, while others will degrade naturally into harmless substances. But many toxic chemicals do persist, and they get into the food chain and sources of drinking water. They enter plants, which are eaten by animals, and all the while the chemicals become more concentrated. At some point, plants, fish and birds start dying and eco-systems will begin to collapse. Humans have less to eat and the food we do have will damage our DNA and make us sick. It might be a slow poison, but it is one that can wreak catastrophic effects on life.

What do we put into the Earth?

We treat the world like a dustbin. We fill our city air with the fumes of cars and lorries. Smoke from industrial plants throws out heavy metals and particulates that travel the world thanks to powerful air currents. Our farms use fertilizers and pesticides with abandon, most of which end up soaking into the ground or getting washed into rivers, damaging the balance of life.

All of this is the flip side to a remarkable rise in our standard of living. Humans have designed and manufactured a staggering number of artificial chemicals to help us live our lives – between 1930 and 2000, the annual production of man-made chemicals increased from 1 million to 400 million tonnes a year. In the past few decades alone, we have invented some 80,000 new chemicals, and although we may encounter them every day, they are mostly invisible.

Take bisphenol A, a suspected hormone mimic: 700,000 tonnes of this is produced every year in the European Union alone, and used for everything from cleaning metals to producing babies' bottles. We apply other chemicals to our skin every day – phthalates and parabenes in cosmetic creams, for example. Yet more are in the objects around us, such as brominated flame retardants on our sofas. Once we have made use of them, all these chemicals will eventually find their way into the wider environment.

Their environmental impact pales, however, when compared to the long-term effects of agricultural and industrial compounds. Take DDT, for example, a persistent bioaccumulative chemical that can stay in the environment for long periods and does not break down easily. This type of chemical can build up in animal tissue and pass up the food chain or to successive generations through the placenta or by suckling.

DDT was introduced as a pesticide around the time of the Second World War, and quickly became a universal weapon in agriculture and public health to fight disease-carrying organisms such as mosquitoes. A few decades later, it was found everywhere in the environment, spread by the natural wind and water currents of the Earth and carried for thousands of miles by migrating birds and fishes. It was detected in the air of cities, in wildlife, even in the Adélie penguins of the Antarctic. It also started to accumulate in the fatty tissues of humans. It is now banned because of its toxic effects on a wide range of living beings all over the world.

Another worrying class of pollutant is the hormone-mimicking endocrine disrupters. 'When you look at the structure of a hormone molecule such as testosterone, they have these circular links of carbon and it's exactly that sort of pattern that's present on a lot of the pesticides like PCBs,' says Andrew Derocher, a professor of biology at the University of Alberta. Over time, animals have evolved ways of dealing with naturally occurring chemicals invading their bodies. But the speed at which man-made chemicals have been invented and then released into the environment has had devastating effects. 'No organism has evolved to deal with these human-produced chemicals,' says Derocher. The result is that they accumulate in top predators such as polar bears and panthers, interfering with their development.

Fertilizers are another major source of pollution. These contain copious amounts of nitrogen and phosphorus to help crops grow, but on a typical farm, a significant fraction of the material applied to fields is not used by the growing crops. Instead, it trickles out into rivers and the sea, where the sudden bounty of nutrients can cause algae to bloom, creating a thick blanket that can quickly choke the water and create a 'dead zone', where oxygen is depleted in the water, killing other plants and driving away any fish or marine mammals.

Effects on wildlife

If pesticides and pollutants manage to collect in high enough concentrations in animals, they can kill them outright. Smaller amounts can reduce fertility or stunt growth.

From a physiological point of view, an animal's biological systems cannot tell the difference between a pollutant molecule and, say, one of its own hormone molecules. As a result, its natural mechanisms for dealing with the invading molecules go into overdrive, and may break down more than just pollutants. The result is an imbalance, and since hormones regulate things like growth and physical development, abnormalities are almost inevitable.

'The more polluted a bear is, the less of an immune response it can generate to an immune challenge,' says Derocher. He blames the problem on polychlorinated biphenyls (PCBs), which persist in the environment decades after they were banned. Dutch scientists carried out research on seals in the

1990s and found that animals that were fed contaminated herring caught off the Netherlands, where European rivers run into the sea, had only half the breeding success of those fed fish caught in the open North Atlantic. The former group also had suppressed immune responses.

In many species of carnivorous birds, such as hawks and eagles, pesticides have accumulated to catastrophic levels. Pesticide use has also been implicated in the collapse of bee populations around the world, in itself a doomsday scenario that could cripple the production of food and crops for ever (see p.80).

The destruction of food webs by pollution makes ecosystem problems even worse. If the plants in a body of polluted water are left uneaten (because the animal population has dropped), they will eventually die and fall to the bottom, decaying and releasing hydrogen sulphide and other gases, thus causing further environmental degradation.

On the ground, or rather in it, pesticides and pollutants have more invisible effects. The structure, fertility and formation of soil depends on the huge numbers of invertebrates (from the tiniest protozoa through nematodes to familiar earthworms) that live in the few inches below the surface of the ground. A single square metre of soil can contain as many as a million arthropods, such as springtails, beetles, millipedes, spiders and ants.

'In places where only a few of these animals are present, the soil is usually of poor structure and contains distinct layers of undecomposed organic matter near the surface,' says Clive Edwards of the Rothamsted Experimental Station in Hertfordshire. 'Soil that contains few invertebrate animals or none still produces crops if it is well tilled and artificially fertilized (although in fact the process of cultivation tends to reduce the number of soil animals). If there were no invertebrate populations in woodland soils, however, the process of soil formation would be very slow or would stop altogether, with drastic ultimate effects on the soil's fertility.'

And it is into this delicate area of soil that modern agriculture deposits its most potent chemical pesticides.

Effects on people

In this global chemical dance, humans are just another animal, albeit at the top of the food chain. 'If we were exposing the population to harmful levels of chemicals that can mimic oestrogen and chemicals that can block the action of hormones, the effect you'd see is an increase in hormone-related cancers – cancer of the breast, the prostate, the testes,' says Gwynne Lyons, a director of Chem Trust, a charity that works to protect humans and wildlife from harmful chemicals. 'And you would expect to see an increase in birth defects, [problems with] the reproductive tract, undescended testes. You might also expect girls to be coming to puberty earlier. All those effects do seem to be happening.'

A large number of the 300 or so man-made chemicals found in humans are those that have been banned for decades – PCBs and other pesticides such as DDT – but Lyons points out that it is not possible to turn off the exposure tap overnight, and new effects on health might be discovered far into the future. Scientists have already found evidence, for example, of a link between Parkinson's disease and long-term and heavy exposure to pesticides – the strongest associations were in people with

Parkinson's who had been exposed to herbicide and insecticide chemicals such as organochlorides and organophosphates.

What can we do about it?

Chemicals have given us the world we live in today, and there is no sense in regretting their use. As with so many things after the profligate 20th century, we just have to be more intelligent and sustainable about deciding which ones we should use. Farmers, for example, could get by with less fertilizer without much reduction in their crop yields. Using fewer long-lived compounds is another step, so that any that do escape into the environment get broken down into harmless products quickly.

As far back as 1967, environmental scientist George M. Woodwell wrote about the need to pay more attention to the accumulation of persistent toxic substances in the ecological cycles of the Earth, pointing out that the size of our planet will not let us hide the problem for ever. 'It affects many elements of society, not only in the necessity for concern about the disposal of wastes but also in the need for a revolution in pest control,' he said. 'What has been learned about the dangers in polluting ecological cycles is ample proof that there is no longer safety in the vastness of the earth.'

OZONE DESTRUCTION

The demise of the ozone layer is the environmental catastrophe that everyone was concerned about before climate change came along to steal its thunder. Their reason for worrying about this delicate band of gas? Less ozone in the atmosphere means less life on Earth, plain and simple.

Concern about the ozone layer was at its most fervent in the late 1980s and early 1990s. Thanks to years of industrial-scale refrigeration, this part of the Earth's atmosphere, a shield against the ravages of radiation from space, was thinning. For some, there was even a faraway but apocalyptic thought: if the ozone layer was to disappear completely, what chance would we have of survival?

What is the ozone layer?

Ozone is not pleasant stuff. A type of oxygen molecule that contains three rather than the usual two atoms of the element, it is an irritating and corrosive gas found throughout our atmosphere. The part we are interested in, the ozone layer, starts between 10 and 16 km (6 and 10 miles) above the surface of the Earth and extends up to 48 km (30 miles). Here, in the stratosphere, ozone is more concentrated than anywhere else, accounting for around three molecules in every 10 million molecules of air. It doesn't sound much, but this layer contains 90 per cent of the atmosphere's ozone and it keeps life going on Earth.

This stratospheric ozone is crucial because it absorbs most (up to 99 per cent) of the harmful ultraviolet radiation coming from the Sun. Without this filter, plants and animals would be damaged by high-energy UV rays. According to NASA, every 1 per cent reduction in the Earth's ozone shield would increase the amount of UV light reaching the lower parts of the atmosphere by 2 per cent.

UV radiation, a high-frequency form of light emitted by our Sun, can be divided into three broad categories based on how energetic it is – labelled A, B and C in ascending order of energy.

UV-C is the most dangerous to life, but it is all screened out by the ozone layer and does not reach the Earth's surface. Artificial UV-C radiation has antimicrobial properties, and is used for sterilization (more on this later). Most UV-B is also absorbed by the ozone layer, but some does reach the surface of the Earth. It can be harmful to human skin and is the main cause of sunburn. This is radiation we actually need, in small doses, because it helps our skin to produce vitamin D, a crucial ingredient in healthy bones and nervous systems. Excessive amounts, though, can lead to genetic damage or skin cancers.

The human body's natural defence against UV-B is to increase the level of brown pigment, called melanin, in the skin. This chemical can absorb UV radiation and dissipate it as heat, reducing its potential for harm. It is also therefore the pigment that produces a suntan.

The lowest-energy radiation from the Sun, UV-A, mostly passes through the ozone layer and reaches the surface of the Earth. Though this radiation can be harmful in very large doses, it is significantly less worrying than the other types of UV.

All three types of UV radiation can damage collagen in the skin, which makes it less supple and look older. But the most dangerous result of the rays is their potential to damage genetic material. The radiation can be absorbed by the DNA molecules in the centre of cells, which can cause breaks in the long chains. If the body's repair mechanisms do not spot the breaks and repair them, they can lead to the death of the cell or even mistakes in the copying process when it reproduces. In some cases, this can lead to aggressive cancers.

UV light is dangerous to organisms other than humans too: UV lamps producing high-energy rays are routinely used to sterilize air and water of bacteria and viruses. The radiation breaks molecular bonds in the DNA of the microorganisms, rendering them unable to reproduce.

The ozone hole

Scientists first became worried about a hole forming in the Earth's protective ozone shield in the 1970s. Decades of evidence pointed to an accumulation of human-produced chemicals that seemed able to break down ozone molecules. These compounds – combinations of chlorine, fluorine, bromine, carbon and hydrogen known as halocarbons, and combinations of chlorine, fluorine and carbon called chloro-fluorocarbons (CFCs) – had been widely used for decades in fire extinguishers, refrigerators, air-conditioners and the manufacture of electronics.

These compounds are very stable, and whenever they escape into the atmosphere, they float to the stratosphere in one piece. When they reach the ozone layer, the molecules are split by energetic UV light and the chlorine atoms begin a chain reaction that ends up tearing apart hundreds of thousands of ozone molecules. This means that lots of incoming UV-B radiation is left unabsorbed and is therefore able to reach the Earth's surface.

Satellite measurements of the ozone in the atmosphere show that these gases are collecting largely at the poles, and that over Antarctica the ozone layer has depleted to such an extent that there is essentially a 'hole' there for several months every year during springtime. Measurements earlier this decade showed that over the South Pole, up to 60 per cent of the total overhead ozone is gone between September and November. The atmosphere above the Arctic also experiences ozone depletion of around 20–25 per cent for short periods between January and April every year. All these reductions in ozone are associated with local increases in UV radiation reaching the Earth's surface.

Is the ozone layer likely to go?

In 1986, after it had been established that CFCs and halocarbons were causing so much damage, international governments got together to sign the Montreal Protocol. This required the 195 signatories to stop production of ozone-depleting gases and to develop ozone-friendly versions instead. It meant that CFCs were replaced with hydrochlorofluorocarbons (HCFCs), which in turn will eventually be replaced with compounds that cause no ozone depletion at all, such as hydrofluorocarbons (HFCs).

So far it seems to have worked, with scientists noting that the removal of ozone-depleting gases has slowed ozone loss in the stratosphere in the past decade. According to the US National Oceanographic and Atmospheric Administration (NOAA), the Montreal Protocol means that the ozone layer has a chance of recovering and is expected to get back to normal in the next 50–100 years.

A report by the Stockholm Resilience Centre in 2009 assessed various environmental threats to civilization and gave the stratospheric ozone layer a relatively clean bill of health. 'The appearance of the Antarctic ozone hole was proof that increased concentrations of anthropogenic ozone-depleting substances, combined with polar stratospheric clouds, had moved the Antarctic stratosphere into a new regime,' it declared. 'Fortunately, because of the actions taken as a result of the Montreal Protocol, we appear to be on the path that will allow us to stay within this boundary.'

But ongoing success depends on several factors, according to David W. Fahey, a physicist at NOAA. Governments will need to continue observing the ozone layer to promptly notice unexpected changes, he argues, and they will also have to ensure that nations adhere to regulations – the phase-out of HCFCs will not be complete until 2030. Scientists also need to keep an eye on new industrial chemicals, in case they have ozone-destroying potential.

There is one other wild card, however. CFCs might have been banned, but these are not the only gases that can deplete ozone – others include the oxides of nitrogen, and also hydroxyl ions (produced when water molecules are split high in the atmosphere). In 2009, scientists at NOAA's Earth System Research Laboratory calculated that nitrous oxide (N_2O) was now the biggest ozone-depleting gas emitted by humans. This gas is a by-product of farming and other industrial processes, and is also used by dentists as an anaesthetic (so-called laughing gas).

NOAA scientists calculated that since it is not going to be phased out any time soon, nitrous oxide will continue to erode the ozone layer for the remainder of the 21st century at the very least.

The ozone layer might be recovering for now, but you can bet it will still face some serious risks in the years to come.

ASTEROID IMPACT

It's a common theme of disaster stories: an object is hurtling towards the Earth from outer space, its path inexorable, the potential damage absolute. If life survives after the impact, it will undoubtedly be drastically diminished.

It has happened before. Geological records show that the Earth has been impacted in the past by some very large objects. Sixty-five million years ago, the dinosaurs (along with more than half the other species on the Earth) were wiped out by a six-mile-wide asteroid that smashed into the area around Mexico. In 1908, 5,000 km² (2,000 sq miles) of forest in the Tunguska region of Siberia was flattened by a 300-foot-wide object from space, which disintegrated 6 kilometres (4 miles) from the ground and released the energy of 1,000 Hiroshima-sized atom bombs.

And in 1994 we got our first chance to watch what happens when two heavenly bodies collide, when fragments of the comet Shoemaker-Levy 9 collided with Jupiter – some of the resulting scars on Jupiter's surface were bigger than the Earth itself.

It will happen again. In late 2004, scientists became worried about a 400 m (1,300 foot) wide asteroid called Apophis, named after an ancient Egyptian spirit of evil and destruction, that seemed to be on course to swing very close to the Earth in 2036. NASA estimated that if it hit the Earth, Apophis would release more than 100,000 times the energy generated in the nuclear blast over Hiroshima. Thousands of square miles would be directly affected by the blast, but the whole of the Earth would feel the effects of the dust released into the atmosphere.

Watching the skies

There are two types of object that could cause trouble for Earth. Comets are balls of ice and dust left over from the formation of planets. They normally lurk on the edges of the solar system but can be dislodged by the gravity of the Sun and get into the paths of planets.

Asteroids are hard lumps of rock thought to be the beginnings of a planet that never quite managed to form between Mars and Jupiter. There are more than a million known asteroids, and an object around half a mile wide hits the Earth every 100,000 years. Objects larger than 6 kilometre (4 miles) wide, which could cause mass extinction, will collide with Earth every 100 million years. Experts agree that we are overdue for a big one.

NASA keeps its eye on more than 900 near-Earth objects (NEOs) to make sure nothing gets too close to the planet. None of these are causing concern at the moment. In any case, nowadays we might stand a fighting chance against the rocks from space. Scientists are designing space missions that could theoretically divert or destroy any incoming asteroids.

The effects of a collision

The Earth is bombarded constantly with cosmic debris, at speeds of more than 16 km/s (10 miles per second). Around 100 tonnes of rubble hits the Earth every day, much of this material burning up in the atmosphere with little result other than the fireworks of a shooting star. Some land as small rocks, meteorites, and end up in the collections of institutions such as the Natural History Museum in London, or the Vatican.

Imagine that a dense clump of rock finds itself attracted by the combined gravitational influences of the Sun and surrounding planets and on a crash course for Earth. As it enters the atmosphere at 16 km/s, its surface would heat up and its outermost layers would begin to evaporate away, turning the incoming object into a fireball. The air around it would expand rapidly, sending shock waves and sonic booms across the world that would flatten buildings and trees hundreds of miles away.

According to a report produced by NASA in 2003 on potential asteroid impacts, any object up to 150 m (500 feet) in diameter would disintegrate in the atmosphere rather than reaching the ground to form a crater, a bit like the situation in Tunguska in 1908. In that case, the immediate area would be showered with debris.

If the asteroid was bigger, a sizeable chunk would hit the Earth's surface. If it landed in the ocean, giant tsunamis would radiate from the impact point and drown coastal cities. A 200 m (700 foot) object landing in the Atlantic Ocean, for example, would be enough to deluge all the coastal cities of the Americas, Europe and Africa.

'Those smaller events occur rather more frequently – they

are talking about a once every several thousand years event,' says Duncan Steel of the University of Salford, a leading authority on asteroids and comets.

As the asteroid hit the ground, it would also throw colossal amounts of dust up into the atmosphere. With a sufficiently large impact, dust would reach the stratosphere. The particles would circulate around the Earth's natural weather systems, and there they would stay for some time, all the while blocking the sunlight from the Earth's surface, causing plants to die and, eventually, the animals that live off them.

Can we stop it?

There are many ideas about how to stop the worst effects of an asteroid on course for the Earth, ranging from giant mirrors floating in space that could vaporize parts of it, to methods that rely more on brute force, such as smashing a rocket into it to deflect it. The traditional Hollywood solution of sending a nuclear bomb to the surface of the asteroid, however, is nowhere in sight.

Deflection methods fall mainly into two categories: kinetic and low-thrust. Kinetic methods are those that provide an instantaneous change of properties within the asteroid – sending a nuclear warhead or some sort of exploding device against it to create a shock wave, for example. Low-thrust methods include painting the surface of the asteroid with reflective or absorbing paint so that its properties are changed by attracting more or less light, thus heating or cooling it.

Whatever method is used, it would only change the path of the asteroid by minute amounts. But even very small adjustments could, over the years, build to big changes in its orbit.

The exact method used would also have to vary depending on the type of asteroid. Some asteroids, known as rubble piles, are loose collections of rocks and ice, so slamming a rocket into one of these would be useless, because the energy of the impact would simply be absorbed, much like the crumple zones in a car.

In this case, a more successful method might involve melting part of the surface of the asteroid by concentrating sunlight. A large solar sail or mirror could reflect the Sun's rays on to the surface of the asteroid and burn part of it away. The jets of gas produced would create a small but constant thrust that could deviate the asteroid into a new orbit.

More traditional solid asteroids provide a range of options. You could place an engine on the surface, for example, which would create a very low thrust and move the asteroid ever so gently over an extended period of time. Or you could hurl a spaceship directly into its path. The idea here is not to physically push the asteroid away but to use the collision to gouge out a hole in the rock. The ejection of material would then push the object in a different direction.

The European Space Agency already has plans to conduct an experiment in deflecting asteroids away from the Earth. Its Don Quixote mission will consist of two spacecraft: *Hidalgo* and *Sancho*. *Hidalgo* will smash into an asteroid at high speed, while *Sancho* will watch the collision and record any shift in the asteroid's trajectory.

Piet Hut of the Institute for Advanced Study in Princeton has championed the idea of a robotic tugboat that could attach itself to an asteroid and push it out of the Earth's path. Based on early warning, provided by ground tracking and orbit prediction, it would be deployed ten years or more before potential impact.

The performance of the tugboat, he says, would depend on the development of a high-performance electric propulsion system called an ion engine. Instead of burning chemicals for fuel, these engines propel a spacecraft forward by ejecting charged particles the other way. The thrust is minuscule – equivalent to the pressure of a piece of paper on your hand – but the engine is extremely efficient and lasts far longer than a conventional rocket engine. Professor Hut calculates that such a spacecraft could be used to deflect NEOs up to half a mile across.

Ion engines would also be crucial for another type of probe, the 'gravity tractor'. Instead of landing on an asteroid, though, the gravity tractor would hover near it, using the slight gravitational attraction between the probe and the NEO to change its path.

Stopping an asteroid will require decades of planning, so that the very slight deflections that can be created will have some effect in keeping the object out of the Earth's way.

What are the chances?

A 90 m (300 foot) object crashes into the planet every 10,000 years, triggering a 100-megaton explosion in the air, greater than the largest H-bomb ever tested. A 900 m (3,000 foot) object scores a direct hit on the planet every 100,000 years, with the force of 10 million Hiroshimas.

Monica Grady, an expert in meteorites at the Open University, says it is a question of when, not if, an NEO collides with Earth. 'Many of the smaller objects break up when they reach the Earth's atmosphere and have no impact. However, an NEO larger than 1 km [wide] will collide with Earth every few hundred thousand years and a NEO larger than 6 km, which could cause mass extinction, will collide with Earth every hundred million years. We are overdue for a big one.'

MEGA TSUNAMI

In the Atlantic Ocean, just off the north-west coast of mainland Africa, there is an island on the verge of falling apart. If it fell into the sea, the resulting waves would wipe out many of the major cities of the world and lead to a catastrophic loss of life. And it could go at any time.

Formed by volcanic activity several million years ago, La Palma in the Canary Islands is a beautiful location, a popular destination for holidaymakers looking for idyllic sunshine or deep gorges filled with a diverse array of animals and plants.

Of all the Canary Islands, La Palma is the most volcanically active, having suffered seven major eruptions since it was occupied by the Spanish in the 15th century. The eruptions themselves are dangerous enough to locals, but there is something else about the island that keeps geologists worried, on behalf of the rest of the world. They believe that a future eruption might dislodge from the island's southern edge a chunk of rock twice the volume of the Isle of Man that is currently under pressure from the gases trapped underneath it.

If this rock fell into the Atlantic Ocean, it would trigger waves that would rise up to over half a mile in height and move at the speed of a jumbo jet. The colossal wall of water, a mega tsunami, would destroy any islands in its path, and when it reached the shores of the US, Europe, South America and Africa, its effects would be catastrophic. Tens of millions of people live on the eastern seaboard of the United States alone – New York City, Boston, Miami and Washington DC would all be under water if La Palma fell apart.

'This would be the biggest natural catastrophe in history,' says Bill McGuire, director of the Aon Benfield Hazard Research Centre at University College London. 'There's a problem with all major natural catastrophes. Because we've never experienced these things we don't think that they're going to happen to us. We just ignore them, but these sorts of events have occurred throughout geological history. They're not going to stop happening just because we're around. La Palma is going to collapse into the North Atlantic. It's not a question of if, it's just a question of when.'

Even worse, La Palma is just one of the volcanic islands that could collapse and cause a catastrophic mega tsunami – there are dozens of others dotted around the world's oceans.

What is a mega tsunami?

A tsunami is the name given to a wave caused by the displacement of a very large volume of water. They are usually caused by underwater earthquakes or volcanoes, but can also result from landslides and asteroid impacts.

Tsunamis travel at more than 800 km/h (500 mph), and in the open ocean, the height of the waves they produce might seem nothing out of the ordinary; perhaps just a slight swell in the normal sea surface. A typical tsunami wave has a wavelength (the distance between successive wave peaks) of more than 200 km (125 miles), compared with around 90 m (300 feet) for a typical wave caused by winds. When a tsunami reaches land, and the water it is in becomes shallow, the wave compresses and slows down, and its height grows enormously.

The first directly observed mega tsunami was recorded in 1958, when a magnitude 7.7 earthquake caused 90 million tonnes of rock to drop into the deep water at the head of Lituya Bay, south-east Alaska. The force of the rock in the water led to a 490 m (1,600 foot) high tsunami wave that inundated the land around the bay, stripping away topsoil, snapping trees and submerging boats. Based on the evidence of the rocks in the area, geologists think that Lituya Bay has suffered several mega tsunamis in the past, as recently as the 19th and early 20th centuries.

These huge, destructive waves have been happening on Earth for millions of years. A mega tsunami is likely to have occurred after the asteroid impact that wiped out the dinosaurs and created the Chicxulub crater in Yucatan, approximately 65 million years ago. A series of huge waves also occurred after the asteroid impact that created the Chesapeake Bay crater, about 35.5 million years ago.

And they happen inland too: in 1980, more than 400 m (1,300 feet) of the top of Mount St Helens became detached and fell into a nearby lake, causing waves that reached a maximum of 260 m (850 feet) above the normal height.

What will happen at La Palma?

When looking for the location of the next mega tsunami, geologists have been examining volcanic islands, which were formed by eruptions and can be torn apart by them as well. The island of La Palma came on to the radar after an eruption of its Cumbre Vieja volcano in 1949, which caused a crack in its western flank. At the same time, a vast chunk of rock – 19 km (12 miles) long with a volume of 500 km³ (120 cubic miles) – dropped 4 m (13 feet) into the surrounding ocean. Scientists think this chunk of rock is still moving, and another eruption could be all that is needed for it to break off entirely.

Computer models of the potential collapse show that the disintegration of the rock would release an amount of energy equal to half a year's consumption of electricity in the US.

Within two minutes of the rock entering the water, the resulting waves would be half a mile high. Around ten minutes later, the tsunami would have travelled 240 km (150 miles). After almost an hour, the wave amplitude would have fallen to 90 m (300 feet) but the surrounding islands would have been submerged, and within a few hours, the waves would reach Africa and the Atlantic coast of Europe – Britain, Spain, Portugal and France would all be hit.

The northern coast of Brazil would also be smashed by 40 m (130 foot) high waves, and around eight hours after the collapse of the volcano, the fast-moving tsunami would begin to reach US cities.

'If you were standing on a beach in what is presently Miami, the very first effects you'd probably see is what we call draw-back,' said Gary McMurty of the University of Hawaii, speaking in a BBC documentary on the potential for a collapse at La Palma. 'The ocean would suddenly just pull away. You'd see a tide, a low tide like you've never seen before in your life. It would be actually spellbinding but in the background you'd be seeing this wall and it'd keep coming at you.'

Harbours would channel the powerful waves miles inland, and the tsunami would lead to the loss of millions of lives and cause billions of dollars' worth of damage to property and land. The long-term effect on the economy would be incalculable.

What can we do about it?

If La Palma fell apart tomorrow, there is not much we could do to stop the spread of the mega tsunami or prevent it from causing damage. Having said that, knowing it was about to happen could allow people to prepare and reduce the impact.

Any collapse of the volcano would occur during a future eruption. This would be preceded by days or even weeks of earthquakes and deformation of the land around the volcano, as the gases and hot lava from the Earth's interior swelled up underneath. Monitoring the volcano and watching out for these signs could give several days' notice of a collapse, crucial time for emergency services to mount a response.

'Eruptions of Cumbre Vieja occur at intervals of decades to a century or so and there may be a number of eruptions before its collapse,' says Simon Day of the Aon Benfield Hazard Research Centre. 'Although the year to year probability of a collapse is therefore low, the resulting tsunami would be a major disaster with indirect effects around the world. Cumbre Vieja needs to be monitored closely for any signs of impending volcanic activity and for the deformation that would precede collapse.'

A tsunami early-warning system already operates in the Pacific Ocean, because of the number of events there, and a similar system for the Atlantic would be required to ensure that problems were detected and information passed on quickly.

If an earthquake was about to occur and scientists thought this could be the one to finally destabilize the Cumbre Vieja's western flank, it would be up to prime ministers and presidents around the world to make the decision whether or not to evacuate people from danger areas. This would be no easy task, but at least they would have been warned. And it would beat having to pick up the pieces of a surprise mega tsunami hitting a bustling New York City.

SUPERVOLCANO

The first thing people thousands of miles away would hear is a series of ground-shaking bangs. Outside, in the direction of the noise, they would make out a dark cloud of ash rising into the upper reaches of the atmosphere. It would not be long before the eruption reached, and devastated, their town too.

They might not realize it, but everything within a few hundred miles of that distant cloud, rising from the distant volcano, would already have been obliterated, burnt or blanketed by dust. In a few short hours, a searing hot wind dense with sharp lumps of rock would smash through their own buildings, setting fire to anything it touched, killing any living thing in its path. That wind would reach the coast beyond their city and superheat the water, initiating a tsunami that would travel across the ocean, carrying the volcano's devastation to the other side of the world. In just the first day, the death and destruction would reach unimaginable levels.

And it would continue. In the following years, billions would die of starvation as crops failed and supplies of fresh water became contaminated. Modern life would stop. And despite the technological mastery humans have of the world, there would be nothing we could do to prevent it.

What is a supervolcano?

There are plenty of volcanoes on Earth capable of big, explosive eruptions and of shutting down entire countries for weeks. Supervolcanoes, though, are something very much out of the ordinary. They are so huge, so potentially devastating, that an eruption would have implications for global civilization, effects comparable to the aftermath of a 900 m (3,000 foot) wide asteroid hitting the Earth: fires, dust clouds, debris and tsunamis. Except that a supervolcano eruption is around five to ten times more likely to occur within the next few thousand years than an asteroid impact.

Geologists estimate that every 100,000 years or so, there should be several eruptions large enough to cause global disaster. There have been none in the time that people have kept records – the giant eruptions at Tambora (1815), Krakatoa (1883) and Pinatubo (1991) all caused major local and climatic problems in the months and years afterwards. But human civilization always remained intact afterwards.

Supervolcanoes would be hundreds of times bigger than these eruptions, however, and their effects much worse. Entire continents would be covered in mud, ash and fire. For interminable years afterwards, the Earth's average temperature would drop as the fine particles ejected by the volcano travelled around the world and blocked out sunlight. Global agriculture would be devasted, food supplies disrupted and billions of people would die of starvation.

At its simplest, a volcano is an opening in the Earth's crust that allows magma, ash and hot gas from the planet's interior to escape. They are normally found at the boundaries between tectonic plates, and erupt when the pressure underneath the cap is too high to be contained. In scientific definition, a super-volcano is something that ejects more than a trillion tonnes of material when it erupts. The last known example, the Toba eruption in Sumatra, happened around 74,000 years ago according to geological records, and everything we know about it and other supervolcanoes has been inferred by scientists from the historical effects of such eruptions on the Earth. Toba was the largest eruption in the past 2 million years, releasing around 2,500 km^3 (600 cubic miles) of material, around twice the volume of Mount Everest. A layer of ash 15 cm thick settled on India and southern China – just 1 cm is enough to ruin crops.

Toba ejected over 300 times more volcanic ash than even the biggest eruption of modern times, Tambora in Indonesia in 1815. Following the latter eruption was the 'year without summer' in which Lord Byron wrote his poem 'Darkness' and Mary Shelley composed *Frankenstein*. Temperatures in the northern hemisphere were depressed for around two years.

Back when Toba erupted, there was nothing in the way of civilization or infrastructure to be damaged, but still the event brought the fledgling human race to the edge of survival. As global temperatures dropped by up to 10°C in some places, the resulting ecological devastation left only a few thousand individuals alive at a time when *Homo sapiens* was first leaving Africa.

And there have been bigger than Toba: the biggest event of all time ejected 9,000 km^3 (2,200 cubic miles) of rock and ash. This was the so-called Fish Canyon Tuff event in Colorado, United States, around 27 million years ago. Super-eruptions are so devastating that some geologists believe they might explain some of the Earth's mass extinctions, such as the one that occurred 250 million years ago in the Permian, when more than 90 per cent of the world's marine species were wiped out after the eruption of the Siberian Traps.

What happens on the Earth? What happens in the sky?

As part of an exercise to raise awareness of the potential dangers of supervolcanoes, the Geological Society of London contemplated what might happen if there was an eruption in the (admittedly highly unlikely) location of just a mile from the Houses of Parliament.

'A super-eruption in Trafalgar Square, London, yielding 300 cubic kilometres of magma would produce enough volcanic deposits to bury all of Greater London to a depth of about 150 metres (nearly 500 feet) thick,' said the working group's report. 'A larger super-eruption (1,000 cubic kilometres) would bury the same area to a depth of 420 metres (almost 1,400 feet). These thicknesses do not include extensive ash-fall deposits, which could cover an area greater than all of Europe.'

The immediate vicinity of the eruption would be destroyed beyond repair, and life would be difficult there as the ash quickly washed into the supplies of fresh water. Mudflows would block rivers and lead to floods.

Beyond that, for tens of thousands of square miles, waves of devastation would arrive in the form of incandescent hurricanes of gas and rock, called pyroclastic flows. These can reach 1,000°C in temperature and travel at speeds approaching those of jet aircraft. 'No living beings caught by a pyroclastic flow survive,' wrote the geologists.

And that is not even the worst of it. 'Globally, most repercussions will come from the effects of the volcanic ash and volcanic gases suddenly released into the atmosphere,' they continued. This cloud of aerosols would reach high into the atmosphere to block out sunlight and absorb water vapour.

'The gases [. . .] commonly include significant amounts of sulphur dioxide, carbon dioxide and chlorine,' said the report. 'Dust and gases injected by an eruption into the stratosphere reflect solar radiation back to space or themselves absorb heat, cooling the lower atmosphere. This fact has led to the concept of "volcanic winter". Silicate dust (tiny ash particles) is thought to be less important, because its residence time in the stratosphere is quite short (only a few weeks to months at most). The main thing causing global cooling after a major eruption is sulphur dioxide gas, which reacts with water to form tiny droplets of sulphuric acid, which remain in the stratosphere for two or three years as an aerosol.'

A super-eruption would devastate human civilization, which depends on trade and food that moves around the world. Ash clouds in the atmosphere would prevent air travel indefinitely and hamper satellite communications.

The Earth's climate would eventually return to normal once the aerosols had disappeared, but given the complexity of the climate, it is hard for scientists to predict how long this would take. Observations of relatively small historic eruptions, such as Krakatoa and Pinatubo, showed that the aerosol levels dropped after a couple of years. 'In principle, putting twice as much aerosol in the stratosphere should double the predicted climatic effect. But climate systems are complex, with important feedback processes. Thus the consequences of very much larger injections cannot be forecast with much confidence,' said the Geological Society of London.'

Recent analysis of ice cores shows that the Toba eruption's aerosol fallout could have lasted up to six years, with global temperatures dropping by 3–5°C. It may not sound like much, but a 4°C cooling sustained over a long enough period can cause a new ice age. According to the Geological Society, however, 'great caution is needed in attributing causes and effects in a system as complex as global climate, and more detailed modelling research is required. Initial computer climate-model runs by scientists at the UK Meteorological Office's Hadley Centre for a Toba-sized eruption suggest Northern Hemisphere temperature drops of 10°C. This would freeze and kill the equatorial rainforests.'

Where will it strike?

There are plenty of contenders for the next supervolcano. The Yellowstone volcano in Wyoming rumbles from time to time, the result of a plume of molten rock that starts deep within the Earth. The area has experienced huge eruptions in the past: 2.1 million years ago, it suffered a Toba-sized blast at the Huckleberry Ridge that created the Island Park caldera and covered most of the continental United States in ash; 1.3 million years ago, a smaller eruption formed in the Henry's Fork caldera.

Other danger spots around the world include Lake Taupo in New Zealand and the Phlegrean Fields volcano west of Naples in Italy. There is also the more familiar region of volcanic activity around Indonesia and the Philippines, while geologists continue to keep an eye on Japan, Central American countries and the Kamchatka peninsula in eastern Russia.

Can we stop it?

In short, no. Neither can we cross our fingers and hope it goes away as an issue. Another supervolcano eruption is inevitable. 'It is not a question of "if" – it is a question of "when",' Bill McGuire, director of the Aon Benfield Hazard Research Centre at University College, London, told *New Scientist* magazine.

Problems such as global warming, impacts by asteroids and comets, rapid use of natural resources and nuclear waste disposal require world leaders and governments to address issues with very long-term consequences for the global community. Countries already have disaster plans in place in case of emergency, and they should put one in place for supervolcanoes too, concluded the working group for the Geological Society of London. What would happen, they asked, if several billion people needed to be evacuated from most of Asia, while at the same time, Europe and North America were threatened with several years of agricultural devastation?

'This is not fanciful, but the kind of acute problem and inevitable consequence of the next super-eruption,' says the Geological Society's working group. 'Sooner or later a super-eruption will happen on Earth and this is an issue that also demands serious attention. While it may in future be possible to deflect asteroids or somehow avoid their impact, even science fiction cannot produce a credible mechanism for averting a super-eruption. The point is worth repeating. No strategies can be envisaged for reducing the power of major volcanic eruptions.'

OXYGEN DEPLETION

The world's plants and animals need oxygen to survive. If for some reason the amount of oxygen in the air or sea dropped, it is not hard to conclude that life would become difficult in its current form. Even if the level fell just a small amount, billions of individual living things would die straight away, and billions more that depended on them in the inter-connected ecosystem would eventually die too.

Oxygen is important for life, but don't take it for granted that the Earth will keep a ready supply available for the living things that cover its surface. In the past, the levels of this crucial gas have fluctuated enough to cause major problems. There was a marked period of deficiency in oceanic oxygen during the Cretaceous period, for example, which was the heyday of the dinosaurs. That deficiency, probably the result of an increase in the activity of undersea volcanism, led to mass extinction of life. As scientists learn more about such phenomena and the geological events that led up to them, they see alarming parallels with the warming world of today. History could be readying to repeat itself.

What is an anoxic event?

Around 93 million years ago, the Earth was going through an unusual episode of volcanic activity. Huge piles of lava collected along the seabed, creating island chains such as the Caribbean. The planet was so hot that there were palm trees in Alaska and large reptiles wandered around northern Canada.

At the same time, the seas were extremely low on oxygen, probably due to the very same volcanic activity. Scientists believe that these huge lumps of lava caused ocean circulation to slow down, so that carbon and oxygen did not move around so quickly. On the sea floor, conditions became highly toxic to life, devoid of the oxygen required to keep things alive.

'Episodes of anoxia, known as oceanic anoxic events (OAEs), have occurred periodically during Earth's history, but none was more severe than that which occurred 93 [million years] ago, during the Cretaceous period,' wrote Timothy J. Bralower of the department of geosciences at Pennsylvania State University, in a 2008 article for *Nature*. 'This OAE caused the extinction of large clams known as inoceramids and tiny protists called foraminifera that lived on the sea floor. Profound changes in ocean circulation also led to the production and preservation of enormous quantities of marine organic material that was subsequently transformed into oil during its burial.'

He also speculated how the volcanic activity of the time might have caused this strange anoxic environment. 'One possibility is that the volcanism seeded the upper ocean with metal micronutrients, increasing phytoplankton production, which in turn led to increased oxygen use during the decay of organic matter. Another, not mutually exclusive, possibility is that a consequence of the global warming stemming from volcanically produced CO_2 was a more stratified ocean, in which oxygen delivery to deep waters became restricted,' wrote Bralower.

So, if heat and gases emanating from the Earth and the sea caused this and several other ancient mass extinctions, could the same killer greenhouse conditions develop once again? Peter D. Ward, a professor in the University of Washington's biology department and an expert on ancient mass extinction events, thinks so.

He argues that during mass extinctions of the past, green and purple sulphur-loving bacteria colonized seas that had been depleted of oxygen but were rich in hydrogen sulphide. Chemical evidence found in bands of stratified rock that mark out mass extinction events in the geological record makes it clear that cataclysmic extinctions in the past – an asteroid hitting the Earth, say – were the exception rather than the rule. 'In most cases, the earth itself appears to have become life's worst enemy in a previously unimagined way,' wrote Ward in a 2006 article for *Scientific American*. 'And current human activities may be putting the biosphere at risk once again.'

He continued: 'Scientists have long known that oxygen levels were lower than today around periods of mass extinction, although the reason was never adequately identified. Large-scale volcanic activity, also associated with most of the mass extinctions, could have raised CO_2 levels in the atmosphere, reducing oxygen and leading to intense global warming – long an alternative theory to the impacts; however, the changes wrought by volcanism could not necessarily explain the massive marine extinctions of the end Permian [period]. Nor could volcanoes account for plant deaths on land, because vegetation would thrive on increased CO_2 and could probably survive the warming.'

But in the oceanic sediments from the latest part of the Permian and Triassic periods, scientists have found chemical clues to an ocean-wide bloom of the bacteria that consumed hydrogen sulphide (H_2S). 'Because these microbes can live only in an oxygen-free environment but need sunlight for their photosynthesis, their presence in strata representing shallow marine settings is itself a marker indicating that even the surface of the oceans at the end of the Permian was without oxygen but was enriched in H_2S.'

Modern oceans contain oxygen in almost equal amounts from top to bottom, because the gas dissolves from the atmosphere at the surface and ocean circulations take it further down. In some places, such as the Black Sea, anoxic conditions exist below a certain level. These conditions are perfect for organisms that can live without oxygen and that pump out H_2S, which also dissolves in seawater. This gas bubbles upwards, eventually meeting the oxygen diffusing downwards from the surface. The place where these two domains meet is called the chemocline, and here the green and purple sulphur-loving bacteria enjoy a supply of H_2S from below and sunlight from above.

'Yet calculations by geoscientists [. . .] have shown that if oxygen levels drop in the oceans, conditions begin to favor the deep-sea anaerobic bacteria, which proliferate and produce greater amounts of hydrogen sulfide,' wrote Ward. 'In their models, if the deepwater H_2S concentrations were to increase beyond a critical threshold during such an interval of oceanic anoxia, then the chemocline separating the H_2S-rich deepwater from oxygenated surface water could have floated up to the top abruptly. The horrific result would be great bubbles of toxic H_2S gas erupting into the atmosphere.'

Calculations show that enough H_2S was produced by such ocean belches at the end of the Permian to create extinctions on both land and sea. Ward adds that further scientific models show that the H_2S would also have attacked the ozone layer, which protects life from the Sun's ultraviolet radiation. 'Evidence that such a disruption of the ozone layer did happen at the end of the Permian exists in fossil spores from Greenland, which display deformities known to result from extended exposure to high UV levels. Today we can also see that underneath "holes" in the ozone shield, especially in the Antarctic, the biomass of phytoplankton rapidly decreases. And if the base of the food chain is destroyed, it is not long until the organisms higher up are in desperate straits as well.'

Scientists reckon that the amount of H_2S entering the atmosphere at the end of the Permian was 2,000 times more than that given off by volcanoes today, killing plants and animals. 'Around the time of multiple mass extinctions, major volcanic events are known to have extruded thousands of square kilometers of lava onto the land or the seafloor. A by-product of this tremendous volcanic outpouring would have been enormous volumes of carbon dioxide and methane entering the atmosphere, which would have caused rapid global warming. During the latest Permian and Triassic as well as in the early Jurassic, middle Cretaceous and late Paleocene, among other periods, the carbon-isotope record confirms that CO_2 concentrations skyrocketed immediately before the start of the extinctions and then stayed high for hundreds of thousands to a few million years,' wrote Ward.

As if that was not bad enough, rising ocean temperatures

meant that it was harder for oxygen to dissolve. This would mean there was even less of the gas in the water, ratcheting up the amount of H_2S. According to Ward, 'Oxygen-breathing ocean life would have been hit first and hardest, whereas the photosynthetic green and purple H_2S-consuming bacteria would have been able to thrive at the surface of the anoxic ocean. As the H_2S gas choked creatures on land and eroded the planet's protective shield, virtually no form of life on the earth was safe.'

What are the possible sources of anoxia today?

There are already hundreds of 'dead zones' in waters around the world, areas where H_2S is battling with oxygen. Most notable are those off the east coast of the US, including some around Chesapeake Bay, the southern coasts of Japan and China, the northern Adriatic and the Scandinavian strait of Kattegat. A 2008 study, published in *Science*, counted 405 dead zones around the world, with the largest being in the Baltic Sea, where oxygen does not reach the bottom of the water for most of the year.

These zones are blamed on runoff of fertilizer from the land. This contains large amounts of nitrogen, which contributes to the blooms of algae in the water. When the algae die, they sink to the bottom of the ocean and are broken down by microbes that consume oxygen in the process. More algae leads to less oxygen in the water, killing other animals and plants, including fish and clams. Once the coast is clear and the situation is anoxic, that's when the oxygen-hating microbes move in and pump up levels of H_2S.

If anoxic conditions have led to mass extinctions before, could the same happen again?

'Although estimates of the rates at which carbon dioxide entered the atmosphere during each of the ancient extinctions are still uncertain, the ultimate levels at which the mass deaths took place are known,' wrote Ward. The so-called 'thermal extinction' at the end of the Paleocene began when atmospheric CO_2 was around 1,000 parts per million (ppm). At the end of the Triassic, it was just above 1,000 ppm.

Today CO_2 is around 390 ppm, so we might seem a long way away from any catastrophe. But our levels of fossil-fuel consumption are still putting around 2 ppm into the atmosphere every year, and this rate is expected to accelerate to 3 ppm as more of the developing world burns oil and coal to power its development.

And it is not just CO_2 that we need to worry about – after a certain amount of warming, huge frozen blocks of methane (some 10,000 gigatonnes of which are sitting on the sea floor) will start to melt and escape into the atmosphere. Methane is 25 times more potent as a greenhouse gas than CO_2, and will accelerate the warming experienced by the Earth.

That means that by the middle or end of the next century, amounts of CO_2 and other greenhouse gases could be approaching the levels required to warm the Earth to the same extent as during previous mass extinctions. The conditions that bring about the beginnings of ocean anoxia may then be in place. 'How soon after that could there be a new greenhouse extinction?' asks Ward. 'That is something our society should never find out.'

GEOMAGNETIC REVERSAL

Compasses have guided people for centuries, helping ships across oceans and travellers across deserts. A tiny sliver of metal that aligns itself to the Earth's magnetic field – it's so simple, so reliable. Well, it is reliable as long as the Earth's magnetic field stays put.

If the Earth's magnetic poles started wandering around, compasses would become useless. If the poles did the unthinkable, and somehow shifted wholesale to the opposite ends of the planet, things would get very confusing indeed.

It is a good job that this does not happen. Well, not too often, anyway. Every few hundred thousand years, the Earth's magnetic poles do indeed start to move around, and eventually switch positions. And getting lost during a reversal is not the only thing you would need to worry about.

The Earth's magnetic field is much more than a way to orient life forms on its surface (though it is very useful for that purpose). It also projects out into space and provides a shield against harmful particles and radiation streaming in from our Sun. If this stuff were to reach the Earth's surface in any great amounts, it would tear life apart. High-energy radiation can rip through DNA and cause irreparable damage to delicate biological cells. Never mind the havoc that this radiation would cause to the world's electrical systems by overloading them and shutting them down, perhaps permanently.

If the Earth's magnetic field was lost for long enough during a reversal, it could be a disaster.

How does the Earth get its poles?

Our planet is a bit like a plum, made of several soft layers of different thicknesses and densities surrounding a small, solid core. The land and oceans sit on the skin of the plum, a relatively thin solid layer of the Earth called the crust. Directly underneath the crust is a vast layer of molten rock, the mantle, which stretches to a depth of almost 3,200 km (2,000 miles). Below that is a region of iron-rich liquid, the outer core, which is constantly in motion around a hotter, solid core made from iron and nickel. The liquid and solid cores are the source of the Earth's magnetic field.

The inner and outer core stay hot (at around 4,000 and 6,000°C respectively) partly due to the energy left over from the formation of the Earth billions of years ago, but mostly because of the energy released by decaying radioactive elements that are present in this broiling mass of material. The outer core keeps moving, at speeds of tens of kilometres per year, and as this metal fluid passes across existing magnetic field lines, electrical currents are induced within it. These in turn generate more magnetic field.

The Earth's magnetic field is often characterized as if there was a huge bar magnet buried inside the planet, with a north and south pole at roughly the corresponding points on the globe and field lines emanating from both sides in the characteristic semicircular pattern familiar from school experiments. Though this is a useful shorthand, the real thing is far more complex and variable, with different intensities and directions at different points on the planet's surface. And it varies over time.

Using mathematical models of the Earth's magnetic field from the past few centuries, geologists have been able to track the pattern of subtle differences in different parts of the world. According to the British Geological Survey (BGS), regions of reversed magnetic flux at the core–mantle boundary have grown over time. 'In these regions the compass points in the opposite direction, in or out of the core, compared to that of surrounding areas. It is the growth in area of such a reversed flux patch under the south Atlantic that is primarily responsible for the decay in the main dipolar field. This reverse patch is also responsible for the minimum in field strength called the South Atlantic Anomaly, centred over north-east Brazil. In this region energetic particles can approach Earth more closely, causing increased radiation risk to low Earth orbit satellites.'

Why do the magnetic poles move?

Every so often, the interplay between the motion of the liquid outer core and the solid inner core in the Earth forces the locations of the poles to move, though no one knows what the exact trigger might be. A magnetic reversal happens when the north pole is transformed into a south pole, and vice versa, an event that last happened almost 780,000 years ago.

There are no complete records of the history of any reversal, so scientists' calculations about how fast a reversal happens are based on evidence in rocks containing imprints of ancient magnetic fields as they were formed, in addition to mathematical simulations. The consensus among geologists seems to be that a full reversal can typically take several thousand years to complete, which is lightning-fast by geological standards – though there is some evidence that they happen even faster than that. A study on 15-million-year-old rocks from Nevada, carried out by geologist Scott Bogue of Occidental College, found evidence that showed possible geomagnetic reversal in just four years.

During a reversal, the geometry of the magnetic field would be much more complex than it is now, and a compass might point in almost any direction, depending on location on the Earth and how the field happened to be changing. 'One of the things that is interesting about reversals is that there is no apparent periodicity to their occurrence,' says the US Geological Survey (USGS). 'Reversals are random events. They can happen as often as every 10,000 years or so, and as infrequently as every 50 million years or more.'

What would happen to life?

In the mid-1980s, scientists began noting coincidences in the timing of geomagnetic reversals and mass extinctions.

'There has been a flurry of papers in the past two years on reversals of the Earth's magnetic field and their possible connection with extraterrestrial catastrophic events,' wrote J.A. Jacobs, an Earth scientist at Cambridge University, in a 1986 edition of *Nature*. 'What has sparked this sudden interest is the reported periodicity of approximately 30 million years in the frequency of reversals, comet/asteroid impacts on the Earth and mass extinctions.'

A year earlier, a University of Chicago geoscientist, David Raup, had examined this seeming set of coincidences, arguing that a definitive link between extinction and magnetic reversal was hard to make. There was probably a connection, he said, between mass extinction and impacts from asteroids, and possibly a tenuous one between asteroid impacts and geomagnetic reversals. 'It is thus not impossible that at least some reversals are caused by comet or asteroid impact and it is in this context that the relationship between periodic intensification of reversal activity and periodic extinction becomes important,' he wrote in *Nature*. 'The discrepancies in analytical results mean that no unequivocal conclusion can be drawn; however there is a possibility that biological extinction, magnetic reversal and large-body impact are linked.'

So much for past extinctions, but would a reversal have negative effects on our modern society? Looking at it from first principles, it is safe to assume that if the magnetic field did disappear, the Earth's surface would be bathed in harmful radiation, damaging us and our electronic equipment. But how bad would it be?

The USGS is not convinced that it would be so severe. 'Reversals happen rather frequently, every million years or so, compared to the occurrence of mass extinctions, every hundred million years or so. In other words, many reversals and, in fact, most reversals, appear to be of no consequence for extinctions.'

The pattern of field lines emanating from the Earth, which buffets against the wind of particles coming from the Sun, is known as the magnetosphere. This does protect us from fast-moving charged particles streaming from the Sun, says the USGS, but so does the atmosphere. 'It is not clear whether or not the radiation that would make it to the Earth's surface during a polarity transition, when the magnetic field is relatively weak, is sufficient to affect evolution, either directly or indirectly, and cause extinctions, such as that of the dinosaurs. But it seems that the radiation is probably insufficient.'

The BGS says that even if the magnetosphere failed, the Earth's atmosphere shields us from high-energy radiation as effectively as a 'concrete layer some 13 feet thick'.

Would the loss of a magnetic field affect other animals? 'Some animals, such as pigeons and whales, may use the Earth's magnetic field for direction finding,' says the BGS. 'Assuming that a reversal takes a number of thousand years, that is, over many generations of each species, each animal may well adapt to the changing magnetic environment, or develop different methods of navigation.'

Which doesn't sound too bad at all. Add in the evidence that our non-human ancestors would have been around during the last big geomagnetic reversal and that they don't appear to have suffered as a result, and it seems as though we should be safe from any major harm during a geomagnetic reversal.

It is worth adding a footnote about some things that our pre-human ancestors would not have had to worry about: they might not have suffered too many physical effects from previous geomagnetic reversals, but who knows how their society or lifestyle might have been affected if they were as dependent as we are on iPhones, electrical grids and satellites, all of which would be damaged irreparably by cosmic and solar radiation if the Earth's magnetic field disappeared temporarily during a reversal. That, though, is a story for another chapter.

SUPERSTORMS

You may have experienced big storms before, but nothing could prepare you for this. The winds, blowing at almost 1,100 km/h, flatten everything in their path. Cars, lorries and trains fly into the air, entire forests are ripped from the ground and streets of hurricane-proof buildings are torn from their foundations.

The extent of the storm is unimaginable, covering an area equivalent to the entire continental US. As boulders are thrown around, coastal areas face tsunamis. When this is over, entire cities across many countries will be razed to the ground.

And that's when the devastation goes global. Because all the while the winds were tearing chunks out of the ground, they were also busy destroying the Earth's ozone layer. In time, with no protection from the harmful rays streaming from our Sun, the Earth will be sterilized of life. This is our world after a hypercane.

How big can a storm be?

A hurricane – or typhoon or cyclone, depending on where you are in the world – is the Earth's way of dumping excess energy from the sea and distributing it around the world and back into space. They occur every year in the warm oceans of the world, including the Atlantic, Caribbean, Indian and western Pacific. In the North Atlantic, for example, the season for hurricanes starts in June and runs until the end of November, usually towards the end of that period, after the seas have been warmed by the Sun for many months.

In its normal state, the air above an ocean is not in thermal equilibrium with the underlying water. This is useful because it allows water to evaporate into the air and carry away some of the energy coming direct from the Sun. Small storms form all the time above tropical waters, thanks to the energy raining down from the Sun.

Hurricanes occur when the levels of energy get higher. If a few thunderstorms, for example, start to rotate around an area of low pressure above the sea, they reinforce each other and the system grows stronger. If the winds rise above 119 km/h (74 mph), the storm system can be called a hurricane. Warm, moist air from the surface of the sea will feed power to the hurricane, and as long as it is above water, the system can continue to grow. Warm air at the centre of the storm moves up and away from the surface, reinforcing the low pressure there and causing even more air to rush in from surrounding high-pressure areas.

A category one hurricane has wind speeds between 119 and 153 km/h (74 and 95 mph) and damage at landfall is minimal. Storm surges as a result of the weather system can reach a few metres, causing flooding if it reaches land. Category five hurricanes, in contrast, can be devastating if they make landfall, with wind speeds above 250 km/h (155 mph) and storm surges greater than seven metres.

Typical hurricanes can last a week and move at up to 32 km/h (20 mph) across an ocean. Once it hits land, a storm system will tend to slow down as its source of energy drops and there is more friction. Hurricane Katrina, the most destructive storm to hit the US, made landfall as a category three storm, though it reached a higher category when it was still above the warm waters of the Gulf of Mexico.

Thanks to climate change, the number and intensity of hurricanes is increasing. In the past century, the surface temperature of the Atlantic has risen by 0.7°C, and a 2005 study published in *Science* showed that the number of category four or five hurricanes around the world had almost doubled over 35 years.

In the same year, Kerry Emanuel, a hurricane expert at the Massachusetts Institute of Technology, found that the intensity and duration of major storms around the US had jumped by around 70 per cent since the 1970s. 'My results suggest that future warming may lead to an upward trend in tropical cyclone destructive potential, and – taking into account an increasing coastal population – a substantial increase in hurricane-related losses in the twenty-first century,' he wrote in *Nature*.

A hypercane, to use Emanuel's words, is a 'runaway hurricane' that can inject large amounts of water and dust into the middle and upper stratosphere, where it will have 'profound effects' on the climate and thus on the ability of life to survive.

In a 1995 paper in the *Journal of Geophysical Research*, Emanuel modelled the conditions required for such a nightmare storm to start. He found that a hypercane could begin if an area of sea just 50 km (30 miles) across was heated to more than 45°C. The resulting 1,100 km/h (700 mph) winds would blow around an eye where the pressure might be as low as 30 kilopascals (normal atmospheric pressure is just over 101 kPa), giving the storm a lengthy lifespan. Conceivably the eye could be hundreds of miles across and the storm itself could stretch for thousands of miles.

For comparison, the largest storm ever recorded in modern times was Typhoon Tip in 1979, where winds blew at around 300 km/h (190 mph) and the central pressure was 87 kPa. The extreme conditions needed to form a hypercane, suggested Emanuel, meant that such storms would probably be limited to tropical regions.

Stratospheric damage

The aftermath of big storms every year makes it abundantly clear that winds, storm surges and floods are devastating to people who suffer them. Hypercanes would be much bigger, and their potential for direct damage therefore greater. But what makes them especially dangerous to our global survival is the damage they do to the upper reaches of the atmosphere, 20 kilometres above our heads.

'The most significant characteristic of hypercanes, from the standpoint of environmental impact, is their ability to inject large amounts of mass in the middle stratosphere,' wrote Emanuel. A ring of air between 5 and 32 km (3 and 20 miles) across, travelling up to an altitude of 20 km (12 miles), would bring with it around 107 kilograms of water per second to the middle stratosphere. Within 20 days, this layer of air would be saturated with water and we would see clouds very high up. Unfortunately, these would certainly not be benign.

For a start, dust and aerosols would cover the Earth, reducing the amount of sunlight getting to the surface. In an odd way, though, that would not be the main problem. The injection of large amounts of water into the stratosphere could have significant consequences for the chemistry of that region. Water molecules would split into a soup of highly reactive free radicals, molecules that have spare electrons (or need one to fill their outer orbits), which can tear other, normally more stable molecules apart. In short, water vapour in the stratosphere would destroy the ozone layer.

The water droplets in the clouds themselves would also catalyze a new set of reactions, activating chlorine that came up with the seawater and deactivating nitrogen oxides. This makes the destruction of ozone even more efficient – and is the mechanism through which the Antarctic ozone hole first appeared.

With the ozone layer depleted, the Earth's surface and all its living things would be at the mercy of ultraviolet rays. Everything on land and in the upper reaches of the oceans would soon die. With no plants and no animals, eventually only very few humans would be left.

Is it likely?

Emanuel's conditions for the formation of hypercanes are considerably warmer (some 10–15°C higher) than any recorded measurement of sea-surface temperature, and even with climate change raising water temperatures around the world, it is unlikely that a tropical sea would reach a sea-surface temperature of 45°C on any sort of regular basis. Instead, a hypercane is likely to be the icing on the cake of another world-ending spectacle: an asteroid strike or massive undersea volcano. Emanuel calculated that a hot spot in the ocean was most likely to occur if a 10 kilometre (6 mile wide) asteroid were to hit a shallow sea. Either that, or if a huge underwater volcano erupted and the water above stayed sufficiently calm long enough for the hypercane to build up.

This could be how the dinosaurs were wiped out. Emanuel believes that the asteroid that smashed into the Yucatan peninsula in Mexico 65 million years ago might have caused a hypercane that resulted in the death of most of the world's species. At the time, there was a shallow sea in the area, which would have been pushed to one side when the asteroid struck. The water would then have rushed back into the hot crater and warmed up to the level needed for a hypercane. Hence the global devastation that resulted.

SPACE

SUN STORMS

Our advanced, interconnected world depends on fast electronic connections between people in different countries, powered by the electricity grid. These two networks, built up and improved over several decades, have brought us the on-demand world that we take for granted. It would take something huge to knock all of that out of action, wouldn't it? Well, our planet orbits the thing that could do just that.

The Sun regularly has Earth-sized storms on its surface that end up ejecting dangerous radiation and particles into space. Mostly these dangerous bits of energy head off into deep space. But what would happen if the Earth got in the way? You could kiss goodbye to the Internet and your electricity supply. Banks and governments would not be able to function. Satellites would be blinded. A storm on the Sun could take us back to the Stone Age.

What is a solar storm?

Without the endless supply of energy raining down from the Sun, there would be no life on Earth. That does not mean, however, that our star is some all-benevolent orb that only gives out goodness.

The Sun, like any other star, is a furious mass of gases with unimaginable energy, which emits radiation of all kinds, ranging from the stuff that plants can convert to sugar through photosynthesis, to high-energy particles and rays that would tear apart anything they came across on Earth.

The Earth, for its part, has a security shield to stop the nasty stuff getting in, while letting through the energy that happens to be beneficial to life. This shield, called the magnetosphere, diverts the worst of the Sun's radiation, preventing it from reaching the delicate molecules of life on the Earth's surface. Normally, all we see of this high-energy radiation are the shimmering Northern Lights, the aurora borealis, and its southern equivalent, the aurora australis.

On occasion, however, the Sun will throw out something more than the usual radiation. Magnetic storms on its surface can end up causing flares, explosions that release in one go as much as a sixth of the Sun's entire output per second. If the storms are particularly strong, they will erupt into coronal mass ejections (CMEs), huge clouds of plasma travelling at 8 million km/h (5 million mph), consisting of energetic electrons and protons with smaller amounts of helium, oxygen and iron.

If the effects of these extreme events were to reach the Earth, the results could be deadly – shutting down power grids, disabling satellites and interfering with electronics. In addition, aircraft flying at high altitudes could be exposed to increased levels of dangerous radiation.

In a typical scenario, a solar flare on the Sun would be accompanied by a burst of electromagnetic radiation (including radio and visible waves in addition to more dangerous gamma, ultraviolet and X-rays) that, when it arrived on Earth, would ionize the outer atmosphere. People on the ground would be safe, but GPS and other satellites would be affected. 'GPS is a critical part of almost everything we do,' says Thomas Bogdan, director of the Space Weather Prediction Centre in Colorado. 'The ubiquitous need for an uninterrupted power supply, satellite-delivered services – every time you go to a gas station and purchase a gallon of gas with your credit card, that's a satellite transaction taking place – and, of course, aviation and communications. We have made our lives increasingly dependent on these things, but each of them carries vulnerabilities to space weather with them.'

Around 10–20 minutes after the initial flare would come a burst of energetic protons. 'Now at risk would be satellites at geostationary orbit – if they do not have sufficient shielding around their sensitive electronics, they could be subject to problems with the internal computational activities,' says Bogdan.

A further 10–30 hours later, a CME would hit the Earth's magnetosphere and cause electric currents to surge along oil pipelines and high-tension electricity lines. This might cause blackouts such as the one that occurred in Quebec in 1989. Around large parts of the world, people would see a lightshow in the sky similar to the aurora borealis.

'Space weather can affect human safety and economies anywhere on our vast wired planet, and blasts of electrically charged gas travelling from the Sun at up to 5 million miles an hour can strike with little warning,' warned John Holdren and John Beddington, respectively the chief scientific advisers to Barack Obama and the UK government, in a joint statement

in 2011. 'Their impact could be big – in the order of $2 tril-
lion during the first year in the United States alone, with a
recovery period of four to ten years.'

History is the guide

The largest solar storm on record occurred in 1859. The British astronomer Richard Carrington noticed a succession of
freak events, including compasses going crazy and aurorae in
the sky as far south as Cuba. There was little electric infrastructure in place around the world at the time, but the solar
storm did send currents running along the newly built tele-
graph systems. 'They were so strong that the operators of
the telegraphs could disconnect their batteries and still start
sending messages,' says Bogdan.

Holdren and Beddington outline further events. 'In 1921,
space weather wiped out communications and generated fires
in the northeastern United States. In March 1989, a geomagnetic storm caused Canada's Hydro-Quebec power grid to
collapse within 90 seconds, leaving millions of people in darkness for up to nine hours. In 2003, two intense storms traveled
from the Sun to Earth in just 19 hours, causing a blackout in
Sweden and affecting satellites, broadcast communications,
airlines and navigation.'

The 1989 storm, in particular, has gone down in history as
an example of what can happen to modern infrastructure. 'On
Friday March 10, 1989, astronomers witnessed a powerful
explosion on the sun. Within minutes, tangled magnetic forces
on the sun had released a billion-ton cloud of gas. It was like
the energy of thousands of nuclear bombs exploding at the
same time,' says Sten Odenwald, an astronomer at NASA. 'The

solar flare that accompanied the outburst immediately caused short-wave radio interference, including the jamming of radio signals from Radio Free Europe into Russia. It was thought that the signals had been jammed by the Kremlin, but it was only the sun acting up.'

On 12 March, the CME finally hit the Earth's magnetic field, causing a huge geomagnetic storm. There were spectacular lights in the sky, but on the ground, the particles were inducing currents in the power grids of North America. 'Just after 2:44 a.m. on March 13, the currents found a weakness in the electrical power grid of Quebec,' says Odenwald. 'In less than 2 minutes, the entire Quebec power grid lost power. During the 12-hour blackout that followed, millions of people suddenly found themselves in dark office buildings and underground pedestrian tunnels, and in stalled elevators. Most people woke up to cold homes for breakfast. The blackout also closed schools and businesses, kept the Montreal Metro shut during the morning rush hour, and closed Dorval Airport.'

Meanwhile, in space, NASA's TDRS-1 satellite spun out of control for several hours, and space shuttle *Discovery* had mysterious sensor problems on the high-pressure tanks that supplied hydrogen to its fuel cells.

Is it likely?

At the 2011 meeting of the American Association for the Advancement of Science in Washington DC, John Beddington made an unequivocal statement: 'This issue of space weather has got to be taken seriously. We've had a relatively quiet [period] in space weather and we can expect that quiet period to end. Over the same time, over that period, the potential vulnerability of our systems has increased dramatically. Whether it's the smart grid in our electricity systems or the ubiquitous use of GPS in just about everything these days.'

Jane Lubchenco, administrator of the US National Oceanic and Atmospheric Administration, agreed. 'It's reasonable to expect there will be more [solar storm] events. The watchwords are predict and prepare.'

Their calls for more action came with a warning. Solar storms can happen at any time, but tend to become more severe and more frequent in roughly 11-year cycles. The peak of the latest cycle occured in 2013, though it was one of the least active for almost a century. Since the last peak in activity, the world's reliance on electronic technology – and therefore its vulnerability to space weather – has increased substantially.

'A study by the Metatech Corporation in 2008 showed that a repeat of the 1921 solar storm today would affect more than 130 million people with sudden and lasting ramifications across the US social and technical infrastructure,' said Holdren and Beddington in their 2011 statement. They added that a recent report by insurance market Lloyd's of London stated that 'A loss of power could lead to a cascade of operational failures that could leave society and the global economy severely disabled.'

What can we do?

Monitoring the activity of the Sun more closely is one part of the equation. The time it takes for the worst bits of a solar storm to travel from the Sun to Earth does give authorities a window of opportunity to get ready for the coming electromagnetic disturbances.

Meanwhile, power companies could prepare by hardening transformers at substations and installing capacitors to soak up current surges that would result in serious problems. Critical satellites should be shielded – though this is something of a balancing act, since every pound of extra weight increases the amount of money and fuel it takes to get the satellite into space.

'Some of these measures can bear fruit quickly, while others will pay off over the longer term,' say Holdren and Beddington. 'What is key now is to identify, test, and begin to deploy the best array of protective measures practicable, in parallel with reaching out to the public with information explaining the risks and the remedies.'

POLAR SHIFT

The Earth's orientation and orbit around the Sun is critical in ensuring that life everywhere gets the energy it needs thrive. But what if that relative position started to drift, and the Earth began to topple over?

As long as we keep moving around the Sun and all get a share of the life-giving energy from our star, as long as the seasons carry on as normal so that plants can grow and provide us with food, oxygen and aesthetic pleasure, we can largely forget about the Earth's specific movements in space.

But when you start thinking, you realize a few important things. The diversity of life on Earth is a result of a long chain of specific circumstances, such as the chemicals available in a particular pond at some point in pre-history, which kick-started the evolutionary process. Outside all of that is one crucial and overriding condition: the position of the Earth relative to the Sun at all times.

The Earth's particular orbit around the Sun is one of the key factors in determining the climate on our planet. For the first organisms to emerge from the primordial soup 4 billion years ago, the distance from the Sun and the tilt of the Earth's axis had to be just right. For the myriad organisms to have subsequently evolved and flourished, the conditions on Earth had to continue to be right. For life to continue, the conditions will have to stay right.

But the Earth will not remain in the same place relative to the Sun for ever. In fact, it is always moving.

How the Sun heats the Earth

Our planet has an almost circular orbit around our star, and the extent to which this orbit departs from a perfect circle is measured by a number known as the 'eccentricity'. This value changes over hundreds of thousands of years as the Earth is buffeted by the gravitational fields of other planets, particularly Jupiter and Saturn. A low eccentricity means an almost circular orbit, whereas a high eccentricity means that the orbit is slightly elliptical, so there is a wider variation throughout the year in the Earth's distance from the Sun and, therefore, the total amount of energy falling on the surface over the course of the year.

The angle of the Earth's rotational axis is the main driver of how any energy that reaches our planet is stored and distributed. Known as the 'obliquity', it is measured as the angle between our planet's axis of rotation and a line perpendicular to its orbital plane around the Sun.

In the early 20th century, the Serbian astronomer and mathematician Milutin Milankovitch proposed a link between the Earth's climate and its slowly moving position and angle relative to the Sun. He worked out that the variations in the orbital characteristics (including eccentricity, precession and obliquity) taken together affected the amount of sunlight hitting the Earth's surface and, over millions of years, the rise and fall of ice ages on our planet.

Over the course of around 41,000 years, Milankovitch calculated, obliquity oscillates naturally between 22.1 and 24.5 degrees; every 100,000 years, the eccentricity of the Earth's orbit ranges from 0 to 5 per cent.

As the Earth's obliquity rises, the summers get warmer and the winters get cooler. This is because, as the planet tilts further towards the Sun, it receives more hours of sunlight and the light it gets will be at an angle nearer the vertical, which means that it will heat the surface more efficiently. Conversely, in winter, higher axial tilt reduces the energy hitting the surface.

At present, the Earth is tilted in the middle of this oscillation, at around 23.4 degrees, and its obliquity is decreasing. In comparison, the axial tilt of Venus is almost 180 degrees, because it rotates in the opposite direction to the other planets, with its north pole pointing in the direction we would call 'down'. Uranus spins on its side relative to Earth, with an axial tilt of around 97 degrees.

Why does the axis change?

Scientists predict that if everything goes as predicted, the Earth's obliquity will continue to decrease until it reaches a minimum value of around 22 degrees in around 10,000 years. During this time, summers will become cooler and winters will warm up. The effects can be dramatic.

When obliquity is at its lowest, for example, the higher latitudes of the Earth, and particularly the poles, get much less solar radiation, and conditions become more favourable to the formation of glaciers.

There is good evidence to suggest that shifts in the Earth's axis and orbit around the Sun have had devastating effects on complex life forms throughout our planet's history.

In 2006, Jan van Dam of Utrecht University examined the fossilized teeth of more than 100 types of rodents in Spain in an attempt to work out how often species of rats and mice there had risen and fallen in the period from 24.5 million to 2.5 million years ago. He found that species extinction happened in two distinct cycles, every 1 million and 2.5 million years.

These cycles correspond to times when two or more of the cycles calculated by Milankovitch peaked together, leaving the Earth much cooler than normal. 'Pulses of turnover occur at minima of the 2.37-million-year eccentricity cycle and nodes of the 1.2-million-year obliquity cycle,' van Dam wrote in *Nature*. This 'astronomical hypothesis' for species turnover provided a crucial 'missing piece in the puzzle of mammal species and genus-level evolution'. The hypothesis also offered a 'plausible explanation for the characteristic duration of 2.5 million years of the mean species lifespan in mammals, and may explain similar durations in other biological groups as well', wrote van Dam.

Extreme changes in obliquity can also have dangerous consequences. In a 2003 study published in the *International Journal of Astrobiology*, Darren M. Williams and David Pollard of Pennsylvania State University took Milankovitch's ideas further by modelling what would happen to life on Earth if the planet's axis became tilted further than its natural limits. Life might not be destroyed completely, they concluded, but advanced civilizations such as ours would be in grave danger of being cooked. Any Earth-like planet with obliquity greater than 54 degrees would experience huge changes in climate, with 'temperatures reaching 80–100°C over the largest middle- and high-latitude continents around the summer solstice', they wrote.

The high-temperature extremes exhibited in most of their simulations would be problematic for 'all but the simplest life forms on Earth today'. 'Photosynthetic organisms would be challenged by the long periods of darkness that would affect nearly an entire hemisphere for months. Some of our planets might only be suitable then to a class of organisms known on Earth as extremophiles, which occupy the dark ocean bottom or deep underground and which can withstand temperatures approaching 400°C, provided they are near a source of water. Such organisms would easily withstand the temperature variations of extraordinary amplitude that we have simulated here.'

Some form of life could survive if the Earth's axis shifted to an extreme obliquity, but it would not be able to live on the continents, where summers would be unbearably hot and water resources would presumably run dry. In this scenario, any human habitation would have to be abandoned.

Could it happen?

Aside from the natural oscillations, could the Earth's axis be forced to change in other ways? Felix Landerer of NASA's Jet Propulsion Laboratory has proposed that global warming has slowed our planet's obliquity, as warming oceans and melting ice sheets shift the amount of physical material sitting on different parts of the Earth's surface. He estimates that the rush of fresh water into the oceans from the melting Greenland ice sheet, for example, is causing the Earth's axis to tilt by just over an inch per year.

Using a computer model, Landerer predicted that a doubling of carbon dioxide in the atmosphere by 2100 (which is only the most moderate of projections by scientists) would push more water on to the Earth's shallower ocean shelves. This redistribution of mass would move the Earth's northern pole by around half an inch per year in the direction of Alaska.

Earthquakes can also shift the Earth's axis. The magnitude nine quake off the north-eastern coast of Japan in March 2011 moved the Earth's axis by around 15 cm (6 inches), according to meteorologist Bethan Harris of the University of Reading. It also moved the Japanese land mass by several feet, and this redistribution means that the Earth's rate of rotation has increased (albeit by a tiny amount).

A more frightening scenario is if our Moon was somehow knocked out of orbit, given that our planet is effectively held upright (at low obliquity) by the strong gravitational influence of our satellite companion. 'Earth's axial tilt is stable with the Moon present for obliquities of less than 60 degrees,' wrote Williams and Pollard in their 2003 paper, which modelled extreme axial tilts on Earth-like planets. 'Without the Moon, Earth's obliquity would vary chaotically as a consequence of solar tides between 0 degrees and 90 degrees on timescales of less than 10 million years. This result [suggests] that the Moon is in some sense necessary for the existence of life on Earth because it stabilizes the spin axis at low obliquity and maintains climatic clemency over most of the planet.'

LETHAL SPACE DUST

Everything in our quiet corner of the galaxy seems far apart, slow-moving and, well, peaceful. But it is all illusion. Just because there have been no cosmic catastrophes in the mere blink of time that coincides with our lifespan, it hardly means that our galactic neighbourhood is eternally safe.

This might sound strange. Yes there are rocks that hurtle around every so often, and many of them have hit the Earth over the course of its 4-billion-year history, but there is nothing too massive in our vicinity, no supermassive stars or weird black holes. Why the worry? To answer this, you need to think in cosmic time scales.

The Earth moves around the Sun, and the Sun is also moving, around the centre of the galaxy. During that 250-million-year orbit, travelling at 200 km/s (125 miles per second), our solar system passes through all manner of clouds of dust and rock, raising sharply the number of asteroids that bombard the planets. When we are passing through the densest regions of space, billions upon billions of lumps of rock rain down on the Earth and the other planets, for thousands of years at a time.

In the 3 billion years that there has been life on Earth, this periodic rain of dust and rock seems to have virtually wiped out all our planet's species several times. In the history of the Earth, cycles of destruction are all too regular.

Regular extinctions, regular impacts

The fossil record shows that around every 30 million years, a large number of species on Earth become extinct. Over roughly the same period, the cycle of impacts from cosmic objects also seems to rise and fall. More asteroids, comets and other debris, which normally sit at a safe distance from the Earth, seem to become attracted to it during these periods. Among all the many inconsequential bits and pieces of dust and ice, there can be something massive enough to cause global problems – boulders a mile across that can cause tsunamis as they hit our surface and block out the Sun by throwing dust into the atmosphere. This cycle of devastation can be traced back for more than 250 million years, and includes the end of the Cretaceous period 65 million years ago, when the dinosaurs became extinct.

What causes this periodic bombardment? When scientists began examining the clues – including factoring in where the asteroids might be coming from and the reasons why they have appeared to be on collision courses with Earth so often throughout history – they came to a startling conclusion. The periodic extinctions were happening, it seemed, because the entire solar system was moving up and down through the plane of the Milky Way at the rate of one cycle every 30 million to 35 million years. During this process, whenever the Earth crossed the densest part of the galactic disc, it got in the way of lots of cosmic objects.

'The approximate 30 [million year] periodicity that appears to exist in the record of mass biological extinctions and terrestrial impact cratering has been interpreted [. . .] as evidence of a quasi-periodic bombardment of the Earth by comets that have been perturbed into Earth-crossing orbits by close gravitational encounters of the Solar System with massive interstellar clouds of gas and dust,' wrote Richard Stothers of the NASA Goddard Space Flight Center in a *Nature* paper in 1984. 'In this model, the underlying clockwork is the Solar System's vertical oscillation through the midplane of the galaxy, towards which most of the massive interstellar clouds are concentrated.'

This vertical oscillation brings the solar system into the path of danger, and also disturbs the huge cloud of dust and rock, called the Oort cloud, that peacefully envelops our solar system. This only contributes further to the catastrophic hail of rocks that smashes into the Earth every 30 million years or so.

The Oort cloud

Most of the asteroids and comets that come near the Earth today originate either in the belt of asteroids that orbits the Sun between Mars and Jupiter, or else in the region of space just beyond Pluto called the Kuiper belt. To most intents and purposes, this is the edge of the solar system, the furthest we have detected objects and known that they were influenced by the gravity of the Sun. But this is not the end of the story.

It might be common to think of the solar system as ending at the orbit of the most distant known planetary objects, such as Neptune and Pluto, but the Sun's gravitational influence extends more than 3,000 times further, halfway to the nearest stars. 'And that space is not empty – it is filled with a giant reservoir of comets, leftover material from the formation of the solar system,' says Paul Weissman, an astronomer who specializes in the dynamics of comets at NASA's Jet Propulsion Laboratory in California. 'That reservoir is called the Oort cloud.'

Weissman refers to the enormous, spherical Oort cloud as the Siberia of the solar system, a 'vast, cold frontier filled with exiles of the Sun's inner empire and only barely under the sway of the central authority. Typical noontime temperatures are a frigid 4°C above absolute zero, and neighbouring comets are typically tens of millions of kilometres apart. The Sun, while still the brightest star in the sky, is only about as bright as Venus in the evening sky on Earth.'

The cloud was named after the Dutch astronomer Jan Oort, and its presence has been inferred, rather than directly observed, from its physical effects – a steady trickle of comets with very long periods (in other words they take a long time to complete one orbit around the Sun) that get into the inner solar system.

In 1950, Oort showed that the comets in this vast cloud were so weakly bound to the Sun by gravity that a random passing star could easily change their orbits. Around a dozen stars pass within one parsec (just over 3.2 light years) of the Sun every million years or so, and this is enough to stir some of the comets into action. Oort described the cloud as a 'garden, gently raked by stellar perturbations'.

Sometimes a star can travel right through the Oort cloud, which causes a violent shake-up of the comets there. Statistically a star is likely to pass within 10,000 astronomical units of the Sun every 36 million years, and within 3,000 astronomical units every 400 million years (an astronomical unit is the distance between the Earth and the Sun). These close encounters do not have any direct effects on the planets of the solar system, but their indirect effects via the comets can be devastating.

A study in 1981 showed that a close passage of a star could send a rain of comets towards the planets, raising the bombardment rates enough to cause mass extinctions on Earth. A few years later, Weissman calculated that the frequency of comets coming into the inner solar system during such an event could reach 300 times the normal rate and last up to 3 million years.

Kenneth Farley, a geochemist at the California Institute of Technology, found observational evidence for Weissman's argument when, in 1998, he examined how much interplanetary dust had been collecting in the Earth's ocean sediments throughout prehistory, a reflection of the number of comets passing by the planets. He found that the influx of comets spiked at the end of the Eocene epoch, around 36 million years ago, a time associated with a moderate biological extinction event. The influx decreased over the following 2 to 3 million years, just as the theoretical models predicted.

Another theory is that the Oort cloud is affected whenever the solar system passes through the galaxy's spiral arms, which contain vast clouds of molecules and massive blue stars. These will also have a gravitational effect on the cloud of comets at the edge of the solar system. These birthplaces of stars and planetary systems can be between 100,000 and a million times more massive than the Sun. Getting close to such a big mass of material would rip comets out of the Oort cloud and throw them in all directions.

Are we at risk?

In an article for *Scientific American* in 1998, Weissman pondered whether the Earth was in any present danger from a shower of comets that had been dislodged from the Oort cloud by stars or molecular clouds. 'Fortunately not,' he concluded. Using positions and velocities measured by satellites, he reconstructed the paths and motions of stars near the solar system. He found evidence that a star has passed close to the Sun in the past million years but that the next close passage will occur in 1.4 million years, a small red dwarf called Gliese 710, which will pass through the outer Oort cloud about 70,000 astronomical units from the Sun.

'At that distance, Gliese 710 might increase the frequency of comet passages through the inner solar system by 50%,' wrote Weissman. 'A sprinkle perhaps, but certainly no shower.'

And what about the massive molecular clouds that the solar system might encounter on its journey around the galaxy? These encounters, though violent, are infrequent, according to Weissman's analysis, occurring only once every 300 million to 500 million years. Over the entire history of the solar system, molecular clouds have had about the same cumulative effect as all passing stars.

But it is worth thinking again about the length of time the Earth and, hopefully, humans will be around. Over cosmic timescales, even the most improbable things are bound to happen.

RUNAWAY BLACK HOLE

Black holes have the ability to induce epic fear. We know they have unimaginable levels of power; we know they can tear apart stars and dust clouds billions of times bigger than the Earth or the Sun. If one happened to get close to our solar system, it would make short work of us before moving on.

The end of our solar system would come in stages. For decades, no one would be able to work out why the number of asteroids hitting the Earth was suddenly shooting up. Astronomers would eventually pick up some wobbles in the orbits of the outer planets, as if a giant invisible hand was knocking them around. Any clouds of dust and gas around the solar system would begin to glow as they became sucked into the black hole and released intense electro-magnetic radiation.

As it moved closer to the solar system, the black hole's immense well of gravity would rip some planets apart and swallow others whole. When it reached Earth, it would draw us into the mysterious void at its centre, a place with a vast appetite for anything the universe can throw at it, a place with infinite destructive potential.

What is a black hole?

Anything can fall into a black hole, but nothing can get out. Stars and planets can disappear and anything that gets too close will be torn apart into its constituent atoms. These cosmic objects are a one-way ticket to mystery, a place where known physics seems to break down and the space we are all familiar with becomes supremely strange.

A black hole is a dead star. After billions of years of shining and fusing hydrogen at its centre, a star will run out of fuel and start to collapse. The collapse increases the temperature and pressure at the centre of the dying star, and the energy levels rise high enough to start fusing the helium there into carbon and oxygen. Later, the helium runs out and the collapse starts again, until the pressure at the core is high enough to begin fusing the carbon.

A star will go through several stages of burning successively heavier fuels before the end of its life: a supermassive star will go through phases fuelled by neon, oxygen and silicon. By the time any star is producing iron at its core, it is near to the end. Fusing iron is no good for a star, since it would consume, rather than release, energy. At this point the star is no longer able to hold itself up against further collapse – from being millions of miles wide, it collapses into a dot (called a singularity) that is smaller than the full stop at the end of this sentence.

Albert Einstein's general theory of relativity predicts that if matter is compressed into a small enough space, the resulting gravity becomes so strong that nothing nearby can escape the pull. The boundary of the region where the gravity of a collapsed star beats every other force around is called the event horizon. Pass this point and there is no coming back, not even for the massless particles of light.

'Once something enters the event horizon, it loses all hope of exiting. Whatever light the falling body gives off is trapped, too, so an outside observer never sees it again,' says Pankaj Joshi, a physicist at the Tata Institute of Fundamental Research in Mumbai. 'It ultimately crashes into the singularity.'

Only the biggest stars collapse into black holes. The Sun, for example, could not naturally become a black hole, because it does not contain enough matter to create the intense gravity needed to overcome the repulsive forces that exist between subatomic particles (the strong and weak nuclear forces and electromagnetism, all of which are usually many orders of magnitude stronger than gravity). Only stars with a mass six times greater than that of the Sun can typically become black holes.

How many are there?

Given that they do not emit light, it is impossible to directly detect or image a black hole. Having said that, scientists infer the presence of these objects based on the effects they have on their local environments.

If anything passes near its event horizon, the black hole's intense gravity will begin to suck in material. A star could end up in orbit, for example, and slowly lose its mass to the hole. As the outer layers are drawn in, the black hole acts like a power plant, releasing gravitational potential energy that energizes intense beams of X-rays and jets of gas. These fly out from the region around the black hole, and are detectable by satellites and instruments on Earth.

Evidence from the past few decades suggests that the biggest black holes are responsible for keeping galaxies together. In 2007, astrophysicists finally confirmed that the Milky Way contained a supermassive black hole at its centre, something that had been suspected for many years beforehand.

There are probably more than 10 million dead stars in the Milky Way that could be candidates for black holes. They are likely to have compressed themselves to point-sized singularities with event horizons around 24 km (15 miles) wide. They are probably cannibalizing everything that wanders into their vicinity right now.

What would happen to Earth near a black hole?

The disaster scenario for our solar system would unfold if any object, such as a planet, got stuck within the event horizon of a black hole. In that region, the gravity would totally dominate, and because of the sharp rate of increase of the forces, different ends of an object would feel different amounts of force. If your head was nearer the hole than your feet, for example, the atoms in your hair would feel a stronger force than those in your toes. This difference would quickly tear you apart, turning you into a spaghetti-like line of atoms moving towards the singularity.

No one knows what actually happens at the singularity, given that nothing, not even information, can escape.

The black hole would not have to sit on top of the Earth, however, for its effects to change the course of human civilization for ever. If it came within a billion miles, its gravitational forces would knock the Earth out of its current orbit and into a dangerous elliptical path around the Sun. In this new orbit, winters would regularly drop to −50°C and summers would reach hundreds of degrees Celsius. It is hard to imagine much life surviving at these extremes.

If the black hole's gravity managed to eject us from the solar system altogether, our planet would end up wandering through deep space without a source of energy to keep us warm or make plants grow. Life on Earth would freeze to death in a matter of months.

Is a black hole headed our way?

Black holes, like any other cosmic object governed by gravity, would orbit the centre of the galaxy and also any other massive objects nearby. Their immense gravitational effects on the objects around them mean that we should notice if they turned up at the solarsystem's edge. But whereas the approach of a star would be obvious, given that it would shine, and astronomers would be able to measure the compression of the oncoming light waves to work out when it would reach the Earth, 'seeing' black holes would be more difficult.

Black holes, as we have said before, do not emit any light. If they happen to eat something on the way to us, we might see a flash of X-rays or be able to detect some superhot gas jets. Beyond that, there would be little warning apart from the effects of the black hole's gravity as it began to reach the outer edges of the solar system. Perhaps a decade or two before the close encounter, the rocks and comets at the furthest reaches of the solar system would start to be thrown around. The black hole might even dislodge an asteroid big enough to cause catastrophic damage if it hit the Earth – though that is a doomsday scenario all by itself.

As the black hole got closer, the shower of asteroids into

the solar system would increase and the outer planets might get knocked out of orbit. By then we would know something untoward was up.

But what could we really do against the colossal cosmic forces and energies that would be unleashed upon us if a black hole came near? Unless we were safely ensconced on another world far away, our planet and people would have to face some horrifying consequences.

Mind you, chance is on our side. Even though there are probably around 10 million black holes in the Milky Way, our galaxy is a vast place that stretches for around 100,000 light years and contains hundreds of billions of stars. We have no idea how many planets (and possibly civilizations) have been destroyed in the vicious maw of a black hole, but it is safe to say that they were the (extremely) unlucky ones.

GAMMA RAYS FROM SPACE

It started with a blinding flash in the sky, a sign that the Earth's atmosphere had been hit by the most intense radiation imaginable. For billions of years, that gamma radiation had zoomed unimpeded through interstellar space at the speed of light. When it crashed into our planet's atmosphere, its immense energy was dumped directly into the air molecules it found, tearing them apart.

The upper atmosphere started to cook. The protective ozone layer disintegrated and all organic matter on the Earth's surface was left exposed to the deadly ultraviolet rays streaming in every day from our Sun. The combination of cosmic gamma rays and local UV rays meant that over the next few months, our planet's surface became largely sterilized of life.

This is what happens when a planet finds itself in the way of one of the biggest explosive events in the universe: the death throes of a supermassive star in the moments before it collapses to become a black hole. As it died, billions of years ago, that star shot two concentrated beams of gamma rays into space, streams of such immense energy that they would easily destroy anything in their path. It was just unlucky that so much time later, our blue-green planet filled with life happened to wander into that firing line.

What is a gamma-ray burst?

Describing anything as 'the biggest in the universe' perhaps seems excessive. After all, the universe is a vast place and we have not yet searched every corner – how do we know something is the biggest? Normally, scientists agree with this sentiment and shy away from such superlatives. But there is one thing that astrophysicists do not hesitate to throw into the category of 'biggest' – gamma-ray bursts. These jets of radiation are the result of the largest explosions known to exist, the final moments of stars at least fifteen times more massive than our Sun.

A star forms when a cloud of hydrogen is drawn together by the mutual gravity of all the atoms. When it becomes sufficiently dense, the gas in the centre will start to fuse and release energy, making the cloud shine. As more gas fuses, the star continues to shine.

Eventually, after several stages of fusion and billions of years of shining, the star will run out of fuel. At this point, it will collapse into a ball of waste, composed of heavy elements that cannot be fused any further. If it is a particularly big star, the subsequent collapse of its insides causes its outer layers to explode into a supernova, an event so bright that it can briefly outshine all the other stars in an entire galaxy.

As if supernovas themselves were not impressive enough, if the star is at the biggest end of the spectrum, the explosion will be so huge that it is called a hypernova. This type of explosion can emit as much energy in just a few seconds as a typical star (our Sun, say) might release in its entire 10-billion-year lifetime.

As part of its explosion, a hypernova sends two concentrated jets of gamma-ray photons shooting off in opposite directions from its poles. This burst of gamma rays, the most energetic electromagnetic radiation there is, can last for anything from a few milliseconds to several minutes. In that time, it will shine about a million trillion times as brightly as the Sun, making it temporarily the brightest source of gamma rays in the observable universe.

According to NASA, the gamma-ray bursts (GRBs) of longest duration originate at the furthest edges of the observable universe, and the stars linked to the explosions are typically in the order of several billion light years away. That means that any gamma-ray photons coming from them would take billions of years to reach us at the speed of light (300,000 km/s). Given that the Earth is just over 4-billion-years-old, it is entirely feasible that some of the GRBs scientists see in the sky today actually happened when our planet was still in its earliest stages of formation, well before life even started to evolve.

Using space telescopes, scientists can detect around one GRB a day from various directions in the universe. Fortunately, all these known events have taken place well outside our galaxy. If a GRB did occur in our galaxy and one of the gamma-ray jets happened to be pointed at Earth, we would be in a lot of trouble.

What would it do to Earth and life here?

If a GRB did have the Earth in its sights, our planet's atmosphere would get pummelled. The incoming gamma rays would likely cause a blinding lightshow as they hit the Earth and knocked electrons off the atoms they encountered. At that point, however, we would not feel any effects on the surface. In the upper atmosphere, the gamma rays would begin splitting nitrogen and oxygen molecules and forcing them to react with each other, causing the creation of toxic brown nitrogen oxide. This is a greenhouse gas that can blot out the Sun and which also also destroys ozone. As soon as the GRB hit the Earth, the ozone layer would be under threat, and in very short order, our planet would be vulnerable to rapid mass extinction.

Charles Jackman, of NASA's Goddard Space Flight Center, has worked out that even a short GRB near the Earth would destroy half of the planet's ozone layer within just a few weeks. Five years later, at least 10 per cent of the ozone would still be missing.

'Nearly all the energy goes into atmospheric chemistry,' wrote physicist Larissa M. Ejzak in a research paper published in the *Astrophysical Journal* in 2006 that modelled the effects of a nearby GRB on the Earth. 'The primary chemical effect of the incident radiation is to break the strong chemical bonds of O_2 and N_2, making possible the formation of molecules that are normally present in very low abundances in the atmosphere. NO and NO_2 are in this class; they also catalyze the destruction of ozone.' She calculated that it would take nearly a decade for the atmosphere to recover from such a burst.

Without the protective ozone shield, harmful ultraviolet rays from the Sun would penetrate to the surface of our planet and start tearing through DNA in living things. For humans, the effects would be gradual: first of all we would notice our skin getting sunburnt more quickly, but underneath, our cells would be quietly ravaged by the UV rays. The rates of skin cancer would skyrocket.

Other animals and plants would suffer as their own cells were unable to reproduce or were killed outright because of the widespread DNA damage. The UV rays would only penetrate the top of the oceans, but this would be enough to kill all the tiny photosynthetic plankton that sit at the root of the oceanic food chain. Remove these and there is much less oxygen being put into the atmosphere and far less food for the animals further underwater.

Perhaps it is a sign of the Earth's resilience that the ozone layer would heal itself after a decade. The bad news is that a decade without phytoplankton would not leave much alive in the oceans.

Is it likely?

Some scientists believe that a gamma-ray burst was behind an infamous mass extinction in which 60 per cent of marine invertebrates were destroyed. 'At least five times in the history of life, the Earth has experienced mass extinctions that eliminated a large percentage of the biota,' says Adrian Melott, a physicist at the University of Kansas who has examined the effects that GRBs might have on the Earth. 'Many possible causes have been documented, and GRBs may also have contributed. The late Ordovician mass extinction approximately 440 million years ago may be at least partly the result of a GRB.'

He added: 'A gamma-ray burst originating within 6,000 light years from Earth would have a devastating effect on life. We don't know exactly when one came, but we're rather sure it did come – and left its mark.'

What is less certain is how often these huge events occur in our vicinity. In a 2004 paper published in the *International Journal of Astrobiology*, Melott made an educated guess that a dangerously close GRB should occur on average two or more times per billion years. So going by statistics alone, we are not due one for at least another 500 million years.

VACUUM DECAY

You would be forgiven for thinking that a vacuum is empty. Indeed, a vacuum is the very definition of empty space. There is nothing there. It is a total absence of stuff. And if there is no stuff, then you might think that there is nothing that could harm us or our world. End of chapter. But you would be wrong.

Among the many strange things that quantum mechanics has revealed about the world is the curious idea that our empty vacuum is not, in actual fact, empty. This conclusion comes from the one bit of quantum mechanics that almost everyone is probably familiar with: Werner Heisenberg's uncertainty principle. This says that it is impossible to know both the exact position and the velocity of a quantum particle, such as a photon or electron, at the same time. The more accurately you know one of these values, the less accurately you can know the other.

Another way to express the uncertainty principle is in terms of the energy and time of the particle, so it is possible that, for extremely short periods of time, a quantum system's energy can be thought of as highly uncertain. In fact, the system can sometimes 'borrow' enough energy from the vacuum to create entirely new particles, as long as those particles do not end up exisiting for very long. These 'virtual particles' appear in pairs (an electron and its antimatter pair, the positron, for example) for a short period and then annihilate each other.

A vacuum, according to quantum mechanics, is not empty at all, but seething with pairs of virtual particles popping into existence and then vanishing. It is a soup of energy, and therein lies danger.

Inflation and false vacuums

To understand why the vacuum is more dangerous than you might at first think, we need to go right back to the start of the universe. Very soon after the Big Bang, after our baby universe had been expanding steadily for a few moments, it started to balloon at an incredible rate. During this period of 'inflation', it more than doubled in size every 10^{-35} seconds; by the time inflation switched off at 10^{-32} seconds after the moment of the Big Bang, this had happened a hundred times. To put that into context, imagine the universe had started off at 1 cm. After 10^{-32} seconds, one 'tick' of inflation, it would be 2.7 cm wide. After two ticks, it would be 7.4 cm. Three ticks later, we're at 20 cm. By 20 ticks, the universe is 4,850 km wide, and after 50 ticks, 5,480 light years. All that in less than 10^{-34} seconds.

By the time inflation had finished, after 100 ticks, the universe would have grown by a factor of 10^{43}. And that is a conservative version of the theory – in some accounts, inflation was even more extreme, with the factor of expansion more like ten multiplied by itself a trillion times.

According to Alan Guth, the physicist at the Massachusetts Institute of Technology who came up with the idea of inflation, this rapid expansion was caused by the release of energy from a form of matter he calls 'false vacuum'. As this decayed into 'true vacuum', it exerted a strange type of repulsive gravity on the space around it, as opposed to the more familiar attractive kind that keeps us stuck to the Earth and the Earth moving around the Sun.

After inflation, things stabilized, and today the universe continues to expand at a much slower, steadier rate. And that is the last we hear of the false vacuum. Well, not quite. 'The false vacuum is unstable, but in most versions of the theory it decays like a radioactive substance, such as radium,' says Guth. This means that the decay is described by a half-life, a time after which half of the false vacuum still remains. After two half-lifes, a quarter of the original vacuum will be left, and so on. That means that, today there is still false vacuum out there somewhere.

As it decays, the false vacuum will expand and the expansion will be faster than the decay. Although only half of the false vacuum will remain after one half-life, it will still be larger than the initial region. 'The false vacuum would never disappear, but instead would continue increasing in volume indefinitely,' says Guth. 'Pieces of the false vacuum region would randomly decay, producing new "bubble" universes at an ever-increasing rate. Our universe would be just one of the universes on this infinite tree of bubbles.'

By sprouting these 'bubble universes', the false vacuum moves into a lower, more stable energy state. Physicists might assume that the same laws of nature would apply in all these individual bubble universes, but Guth is not so sure. 'Other kinds of space might not be three-dimensional, and they might alter the masses of elementary particles, or the forces that govern their behaviour. If there are many kinds of space, the infinite tree of bubble universes would sample all the possibilities.'

How can a vacuum be so dangerous?

So, we have established that a vacuum is more complex than we might have imagined: it can cascade from a higher-energy (false vacuum) to a lower-energy (true vacuum) state by creating bubble universes that expand at the speed of light, with each universe potentially having its own laws of physics.

What if the universe we live in today is in a region of space where the vacuum is stuck in an unstable, high-energy state? In other words, what if there is an even more stable form of vacuum than the one we exist in? It would be as if we are perched at the top of a hill – all it takes is for a nudge to send us tumbling into the lower-energy position at the foot of the hill. The question is, is our vacuum at the top or the bottom of that hill?

The Earth, our solar system, our Sun, our entire galaxy might be in such a false vacuum state right now. At any point, it could collapse into a lower-energy vacuum by creating a new bubble universe. This collapse would grow at the speed of light and rewrite the physics that we know. Under these new laws, the fundamental forces would have different strengths, and our atoms would not hold together in the ensuing wave of intense energy. We, and everything around us, would be torn apart and turned into energy. All that energy might recondense at some point into something else, new forms of matter governed by new laws of nature. But we would not be here to see any of it.

261

In 1980, the Harvard physicist Sidney Coleman calculated that vacuum decay would be the end for all life as we know it. 'The possibility that we are living in a false vacuum has never been a cheering one to contemplate,' he wrote. 'Vacuum decay is the ultimate ecological catastrophe; in the new vacuum there are new constants of nature; after vacuum decay, not only is life as we know it impossible, so is chemistry as we know it. However, one could always draw stoic comfort from the possibility that perhaps in the course of time the new vacuum would sustain, if not life as we know it, at least some structures capable of knowing joy. This possibility has now been eliminated.'

How likely is it?

The theory that underpins vacuum decay is scientifically robust. Whether it would happen in real life is, to all intents and purposes, unknown and unpredictable. The fact that the universe we can see has existed for as long as it has suggests that a bubble nucleation of some vacuum decay has not happened, but that is no guarantee of safety for the future.

If that conclusion leaves you depressed or worried, there is one crumb of comfort you can take. Again, it comes from the same branch of physics that gave us vacuum decay: quantum mechanics.

This rule says that predicting the behaviour of any quantum system is impossible; instead, its equations just provide a range of possible scenarios for each system and assign it a probability of happening. In the 'many worlds' interpretation of quantum mechanics, each of these possibilities actually corresponds to a different universe. For example, if you throw a six-sided die, each of the six possible ways it could land is represented by a different universe. When it lands on, say, a four, you and the entire arena of action move into one of those six universes. In the other five possible universes, which do exist somewhere, the die landed on one of the other numbers.

The same rules of quantum mechanics that might one day lead to catastrophic vacuum decay also lead to the inevitable conclusion that, at the time of decay, new universes will be created that will be spared the destruction. On the one hand, quantum mechanics will take away our lives in an instant. On the other, everything will carry on just as normal.

SOLAR COLLISION

Suppose our parent star was about to suffer a terrible attack. That a cosmic missile was zooming towards our solar system and making for the heart of the Sun. If that missile was a white dwarf – a dense star in its dying days containing the mass of the Sun in a volume one hundredth the size – it would mean the end of the solar system.

As it approached, the white dwarf would alter the orbits of the planets, perhaps pulling one of them (maybe the Earth) into an eccentric path around the Sun or even knocking it out of the solar system altogether. Closer still, it would force a change in the shape of the Sun, elongating it as the missile's intense gravitational field pulled solar gas towards it.

The mutual gravitational attraction would accelerate both stars, and the white dwarf would smack into the Sun at almost 650 km/s (400 miles per second). Anyone watching from Earth would be treated to quite a fireworks display while they managed to cling on to life.

The collision would create a shock wave in the Sun that would compress and heat the entire star above the temperature needed to fuse the hydrogen there. Until that moment, only the centre of the Sun would have been hot enough to fuse anything. In the next hour, the superheated star would release as much energy from fusing atoms as it would have done in 100 million years of normal burning. Releasing energy this fast would cause the gas to expand more quickly than escape velocity – the Sun would have blown itself apart. On the Earth, all our oceans and atmosphere would be obliterated by the rising amount of radiation and the searing gas clouds from the Sun.

An hour after it entered the Sun, the white dwarf would emerge from the other side of our star, almost unchanged and on its way into deep space after destroying all life on Earth.

In space, collisions are common

The idea that a white dwarf is about to smack into our Sun without any warning sounds fanciful. And you can rest easy in that it is highly unlikely: certain parts of the galaxy might be a hotbed of collisions, but estimates of the risk of anything hitting our Sun are around once in 10 trillion trillion years.

For much of the 20th century, though, the whole idea that stellar collisions might even be possible or worthy of study seemed ludicrous to astronomers. According to Michael Shara, curator and chair of the department of astrophysics at the American Museum of Natural History in New York City, 'The distances between stars in the neighborhood of the sun are just too vast for them to bump into one another. Other calamities will befall the sun (and Earth) in the distant future, but a collision with a nearby star is not likely to be one of them. In fact, simple calculations carried out early in the 20th century by British astrophysicist James Jeans suggested that not a single one of the 100 billion stars in the disk of our galaxy has ever run into another star.'

The initial clues that stellar collisions were indeed happening in deep space came with the observation in the 1950s of strange blue stars sitting in the middle of certain globular clusters, which are regions of space dense with stars and dust. Where we are in the galaxy, there is around one star in every ten cubic light years of space; in a globular cluster, the same amount of space might hold hundreds of stars.

Blue stars are among the hottest of all stellar objects, and they burn through their hydrogen fuel much faster than smaller, yellower stars. The globular clusters in which they were situated, however, were known to have exhausted their clouds of gas, which normally give rise to new stars, billions of years earlier. The blue stars that astronomers were seeing were simply too young to have been formed where they were.

Measurements from the Hubble Space Telescope showed that these 'blue straggler' stars were not, as some had thought, normal stars that had somehow conserved their fuel and managed to live longer. Instead, they were two much older stars that had collided and coalesced, creating a brand new star that simply looked a whole lot younger.

When stars collide . . .

What happens when two stars collide depends on a number of things, not least the speed of the objects, what they are made of and how full-on their collision is. 'Some incidents are fender benders, some are total wrecks and some fall in between,' says Shara. 'Higher-velocity and head-on collisions are the best at converting kinetic energy into heat and pressure, making for a total wreck.'

Shara has modelled stellar collisions to work out their effects on the stars themselves and also on the objects around them. He first studied the possible results of a head-on collision between a Sun-like star and a vastly more dense star, such as a white dwarf, several decades ago. 'Whereas the sunlike star is annihilated, the white dwarf, being 10 million times as dense, gets away with only a mild warming of its outermost layers,' he wrote in a 2004 essay for *Scientific American*. 'Except for an anomalously high surface abundance of nitrogen, the white dwarf should appear unchanged.'

If the two stars happened to be of a similar size and density, something different happens. 'As the initially spherical stars increasingly overlap, they compress and distort each other into half-moon shapes. Temperatures and densities never climb high enough to ignite disruptive thermonuclear burning. As a small percent of the total mass squirts out perpendicular to the direction of stellar motion, the rest mixes together. Within an hour, the two stars have fused into one.'

Globular clusters have more collisions than normal regions of space because the stars they contain move relatively slowly. While the Sun is travelling at around 76,000 km/h (47,000 mph) with respect to its cosmic neighbours, the stars in a globular cluster move at around half that speed past each other. This gives gravity some time to work between the objects, deflecting their paths and increasing the chances of a collision. 'The stars are transformed from ballistic missiles with a preset flight path into guided missiles that home in on their target. A collision becomes up to 10,000 times more likely. In fact, half the stars in the central regions of some globular clusters have probably undergone one or more collisions over the past 13 billion years,' says Shara.

Is it likely that a collision will destroy us?

Astronomers are almost certain that collisions between stars must occur, and have uncovered the remnants of those collisions in faraway globular clusters. What they have never done, though, is manage to see one in action. 'To be sure, all this evidence is circumstantial,' says Shara. 'Definitive proof is harder to come by.'

Detecting these faraway collisions will be tough, and might have to be done by looking for gravitational rather than light waves coming from them. If two supermassive objects collide, the event would create a disturbance in space–time. Albert Einstein's theory of general relativity says that any such collision would cause gravitational waves to propagate through space. These waves stretch and move space itself, making the distance between two points expand and contract as the energy passes by.

'The average time between collisions in the 150 globular clusters of the Milky Way is about 10,000 years; in the rest of our galaxy it is billions of years,' says Shara. 'Only if we are extraordinarily lucky will a direct collision occur close enough – say, within a few million light-years – to permit today's astronomers to witness it with present technology.'

Detecting a real-life collision would be exciting for astronomers, giving them an unprecedented insight into the mechanics of stars. But it would also be a stark reminder of just how dangerous such collisions can be.

SCIENTISTS CREATE A BLACK HOLE

As scientists gathered in Geneva to switch on the Large Hadron Collider (LHC) at Cern in 2009, not everyone was celebrating. A handful of people had been sounding an alarm that this gigantic machine might rip apart space itself, creating black holes under the mountain in Geneva and swallowing all of us (and the Earth) into nothingness.

The LHC is the most intricate and sophisticated machine built by humans, its intended use to bring scientists closer than ever before to understanding the Big Bang. It stands as a testament to human ingenuity, a physical embodiment of two decades of design and building work. There is little doubt that it will bring us a better understanding of the most fundamental physics humans have ever known.

But the detractors are not placated by such promises. They were instead focused on one of the scary-sounding scenarios that the particle physicists themselves had predicted: the creation of black holes on Earth.

Black holes of different types

Think 'black hole' and you probably conjure up an image of

something far away in the darkness of space, using its immense gravity to suck in and destroy anything that has the misfortune to wander nearby – stars, planets or spaceships.

This idea is largely correct. Cosmic black holes form in the dying moments of supermassive stars in a process whereby all of what a star once was is crushed by gravity into a dimensionless point of infinite density called a singularity. This implosion is so violent that it tears apart space itself, and within a certain radius of the singularity, nothing can escape. That includes light, hence why black holes are black. We know nothing about what happens inside this radius, the event horizon, because no information can escape the grip of the gravity there.

The size of a black hole depends on its mass – you would have to compress the matter in the Sun to about 3 kilometres (2 miles) across, 4 millionths of its present size, to make it collapse into a black hole; the Earth would need to be squeezed into a radius of 9 millimetres, about a billionth of its present size. Neither of these celestial bodies could turn into black holes naturally, however – there just isn't enough gravity present to crush the masses down.

But cosmic monsters are not the only type of black hole that physicists think is possible in nature. In the 1970s, astrophysicists Stephen Hawking and Bernard Carr looked at whether there might have been black holes in the early universe, when the energies were fierce and the density of matter was immense enough to collapse some regions into tiny singularities. The laws of physics allow for matter to have a density of up to the so-called Planck value of 10^{97} kg per cubic metre, at which point gravity becomes so strong that the material will collapse into a black hole 10^{-35} metres across, with a mass of 10^{-8} kg. Hawking and Carr called these theoretical singularities 'primordial black holes'.

Creating these tiny black holes in the early universe, with its unimaginable density and temperature, is one thing. Producing anything similar on Earth would require an attempt to re-create the conditions of the early universe, in the moments after the Big Bang.

Black holes on Earth

Re-creating the moments after the Big Bang is exactly what the LHC is designed to do. By accelerating protons towards each other at nearly the speed of light, scientists hope to find evidence of new particles in the debris of the collisions and to push back the boundaries of what they know about nature.

When a proton is accelerated in the LHC, it will reach an energy of around seven tera-electron-volts. According to Albert Einstein's equation $E=mc^2$, a proton with this much energy is equivalent to a mass of 10^{-23} kg, around 7,000 times more massive than a proton at rest. 'When two such particles collide at close range, their energy is concentrated into a tiny region of space. So one might guess that, once in a while, the colliding particles will get close enough to form a black hole,' says Carr.

At first glance, there is a big problem with this argument. Two protons with masses of 10^{-23} kg constitutes far too small an amount of material to create Hawking and Carr's primordial black holes, which are a minimum of 10^{-8} kg. Two protons could be made to form a primordial black hole if they were accelerated to much greater velocities, but to satisfy Hawking and Carr's criteria, the particle accelerator needed would have to be the size of our galaxy.

So why is anyone worried about black holes on Earth? The reason is that Hawking and Carr's theory relies on standard general relativity, the mathematical description of gravity formulated by Einstein in the early 20th century. More recent ideas of how gravity might work have proposed that the required density to form a tiny black hole might be much lower than the two scientists had originally speculated.

'String theory, one of the leading contenders for a quantum theory of gravity, predicts that space has dimensions beyond the usual three,' says Carr. 'Gravity, unlike other forces, should propagate into these dimensions and, as a result, grow unexpectedly stronger at short distances. In three dimensions, the force of gravity quadruples as you halve the distance between two objects. But in nine dimensions, gravity would get 256 times as strong. This effect can be quite important if the extra dimensions of space are sufficiently large, and it has been widely investigated in the past few years. There are also other configurations of extra dimensions, known as warped compactifications, that have the same gravity-magnifying effect and may be even more likely to occur if string theory is correct; these have been extensively studied in recent years.'

If gravity really does extend into other dimensions in this way, it would mean that the threshold to form a black hole is much lower than the Planck density. And this in turn means that the density required to make tiny black holes on Earth would lie within the range of the LHC. In 2001, scientists calculated that the lowest boundaries for the Planck density meant that the LHC would produce a black hole every second it carried out proton collisions. They concluded that the particle accelerator at Cern would be a factory for making black holes.

Do we need to worry?

Given the violent nature of the black holes in space, it might be less than comforting to think that similar objects will be popping into existence thousands of times every day at Cern. Could these objects burrow their way out of the collider and begin to eat away at our planet?

This worry is not purely hypothetical, by the way. People become anxious every time a big particle accelerator goes online; before the LHC was due to be switched on, two residents of Hawaii filed a federal lawsuit in an attempt to prevent scientists from going ahead with their work until the potentially catastrophic effects of the particle collisions had been reassessed.

But calm your fears, Frank Wilczek, a physicist at the Massachusetts Institute of Technology, points out that a black hole in space is a very different thing to anything we could create on Earth. He compares the problem to having a single word for all the animals in the world and having elephants in mind when originally defining that word. Amoebas, however, are animals too, he reminds us.

The first thing to remember about any man-made black holes in the LHC is that they would be so tiny that their gravitational fields would not extend very far. A black hole capable of consuming the Earth would have to have a mass of several tonnes – anything generated by the LHC would not even reach an appreciable fraction of a gram.

In any case, the black hole would evaporate and disappear far too quickly to cause any damage. This conclusion comes from the predictions of Stephen Hawking in the 1970s. The idea that black holes could be small led the physicist to wonder whether quantum mechanics, the physics used to describe the smallest constituents of matter, would have any important effects on their behaviour . 'In 1974 he came to his famous conclusion that black holes do not just swallow particles but also spit them out,' says Carr. 'Hawking predicted that a hole radiates thermally like a hot coal, with a temperature inversely proportional to its mass. For a solar-mass hole, the temperature is around a millionth of a kelvin, which is completely negligible in today's universe. But for a black hole of 10^{12} kilograms, which is about the mass of a mountain, it is 10^{12} kelvins – hot enough to emit both massless particles, such as photons, and massive ones, such as electrons and positrons.'

This emission carries off energy, which means that the mass of the black hole would steadily decrease over time. As it shrank, it would get hotter, emitting more particles and shrinking even faster. 'When the hole shrivels to a mass of about 10^6 kilograms, the game is up: within a second, it explodes with the energy of a million-megaton nuclear bomb,' says Carr. 'The total time for a black hole to evaporate away is proportional to the cube of its initial mass. For a solar-mass hole, the lifetime is an unobservably long 10^{64} years. For a 10^{12}-kilogram one, it is 10^{10} years – about the present age of the universe. Hence, any primordial black holes of this mass would be completing their evaporation and exploding right now. Any smaller ones would have evaporated during an earlier cosmological epoch.'

If a proton–proton collision at the LHC was to create a black hole, it would disappear in a flash of Hawking radiation in less than 10^{-26} seconds.

What are the chances?

A study carried out by scientists for Cern pointed out that, although the LHC would achieve energies that no other particle accelerators had previously reached, nature routinely collides particles at much higher energies. Whatever the LHC will do, said officials in a statement reviewing several safety studies of the accelerator, nature has already done many times over during the lifetime of the Earth and other astronomical bodies.

'Cosmic rays are particles produced in outer space, some of which are accelerated to energies far exceeding those of the LHC,' said Cern. 'The energy and the rate at which they reach the Earth's atmosphere have been measured in experiments for some seventy years. Over the past billions of years, Nature has already generated on Earth as many collisions as about a million LHC experiments – and the planet still exists. Astronomers observe an enormous number of larger astronomical bodies throughout the Universe, all of which are also struck by cosmic rays. The Universe as a whole conducts more than 10 trillon LHC-like experiments per second. The possibility of any dangerous consequences contradicts what astronomers see – stars and galaxies still exist.'

The same calculations that show that the LHC will become a factory for mini black holes also predict that around 100 black holes are created every year in the Earth's atmosphere because of cosmic-ray collisions. Over and above everything else, perhaps our own continuing existence is the best evidence there is to reassure us that the LHC is unlikely to create any monstrous black holes that could eat the Earth.

HOSTILE EXTRATERRESTRIALS

What would happen if we ever made contact with aliens, and they decided to come and visit? They might be benevolent, but there is no guarantee. They might just see Earth, with its rich array of elements and resources, as a filling station to be plundered on their way to more interesting places.

When the aliens in science-fiction stories are bad, they tend to be very, very bad. In *Independence Day*, they lay waste to all the world's cities. In *War of the Worlds*, they kill everyone they can find. And it is hard to forget the mischievous aliens in *Mars Attacks!* – blowing up buildings and people with uncontained glee.

No doubt any real-life visitors would be from an advanced civilization, in order to have mastered interstellar travel, and they would probably have the technology and power to do whatever they pleased with us and our planet. If we were lucky, they might not perceive us as a threat, and would ignore us during their visit. Then again, they might leave behind toxic waste or inadvertently wipe us out by bringing a virus or other pest to our planet.

The eminent physicist Stephen Hawking is one of those who is worried. 'If aliens visit us, the outcome would be much as when Columbus landed in America, which didn't turn out well for the Native Americans,' he said in a 2010 documentary made for the Discovery Channel. He argued that instead of trying to find and communicate with life in the cosmos, humans might be best off doing everything they can to avoid contact. When someone as smart as Hawking makes a statement like that, it is worth listening, isn't it?

Is it a bad idea to look for ET?

Astrobiologists and astronomers have been on the hunt for extraterrestrial life for more than half a century, part of a human desire to know whether we are alone in the universe, and to learn from and meet civilizations that are more advanced than our own.

There are billions and billions of stars in our galaxy, and, it is reasonable to expect, an even greater number of planets orbiting them. Some of these, surely, will exist in the 'Goldilocks zone' of perfect temperature and distance from their star (in the same way the Earth exists with the Sun), which would make life possible? And going on sheer numbers alone, it is not unreasonable to expect that some of this life is intelligent and capable of interstellar communication.

The Earth's biggest and most active hunt for alien life started in 1960, when the astronomer Frank Drake pointed the Green Bank radio telescope in West Virginia towards the star Tau Ceti. He was looking for anomalous radio signals that could be signals sent by intelligent life. Eventually his idea turned into the Search for Extra Terrestrial Intelligence (SETI), which used the downtime on radar telescopes around the world to scour the sky for telltale alien signals. For the past fifty years,

SETI has continued its quest, but in all that time, the sky has remained silent.

There are lots of practical problems involved in hunting for aliens. Chief among them is distance. Our galaxy is big – it would take a beam of light 100,000 years to cross from one end to the other. If our nearest neighbours were life forms on the forest moon of Endor, 1,000 light years away, it would take a millennium for us to receive any message that they might send. A response would take the same amount of time to reach the aliens. It is not a timescale that allows for quick banter.

And they might not be communicating in our direction anyway. If the Endorians were watching us, the light reaching them at this very moment from Earth would show them our planet as it was 1,000 years ago. In Europe, that means lots of fighting between knights around castles, and in North America, small bands of natives living on the great plains. If our nearest aliens are tens of thousands of light years away, they would see only the ancestors of modern humans living among much greater beasts. You could forgive them for not bothering to get in touch.

How dangerous could aliens be?

Answering this question in any definitive way is impossible, at least until an extraterrestrial species lands on Earth and makes its intentions clear.

According to Jack Cohen and Ian Stewart of the Mathematics Institute at the University of Warwick, aliens would not look like the canonical 'little green men'. In a commentary in *Nature*, they wrote how extraterrestrial life forms 'might look exactly like people. Or cats. Or houseflies. Or they are invisible, or lurking just outside our space–time continuum along a fifth dimension.'

The lack of a signal from ET has not prevented astronomers and biologists from coming up with a whole range of ideas about what aliens might look like. In the early days of SETI, astronomers focused on the search for planets just like ours. Their idea was that, since the only biology we know about is our own, we might as well assume that aliens are going to be something like us. But there's no reason why that should be the case.

Humans evolved on a planet rich in oxygen and water, where a carbon-based molecule called DNA became the copying mechanism for all life. From our point of view, we seem to exist in a world with just the right parameters of temperature, water and nutrients.

Aliens, of course, are not restricted to our point of view. You don't need to step off the Earth to find life that is radically different from our common experience of it. Extremophiles are species that can survive in places that would quickly kill humans and other 'normal' life forms. These single-celled creatures have been found in boiling-hot vents of water thrusting through the ocean floor, or at temperatures that are well below the freezing point of water. The front ends of some creatures that live near deep-sea vents are 200°C warmer than their back ends.

'In our naïve and parochial way, we have named these things extremophiles, which shows prejudice – we're normal, everything else is extreme,' says Stewart. 'From the point of view of a creature that lives in boiling water, we're extreme because we live in much milder temperatures. We're at least as extreme compared to them as they are compared to us. Similarly for the ones in very cold water.'

He calls this anthropocentric view of life – that whatever is right for us sets the agenda for everything else – the 'Goldilocks mistake'. The problem with the original fairy story, he says, is that even though Goldilocks found baby bear's porridge to her liking, daddy bear was perfectly happy with his hot porridge and mummy bear was happy with her cold porridge. Not to mention the whole forest outside the house that probably didn't even like porridge.

On Earth, life exists in water and on land, but on a giant gas planet, it might exist high in the atmosphere, trapping nutrients from the air swirling around it. 'Most aliens would not wish to visit Earth at all, any more than we would care for a ramble across the surface of a neutron star, or to live, as do some extremophiles, in boiling water,' wrote Stewart and Cohen in *Nature*.

Faced with such diversity, it is still possible to make an educated guess at what aliens might be like. For a start, says Stewart, you have to carve up biological features into things that might be universal across all life forms in the galaxy, and those which are parochial to the Earth.

Parochial features include anything unique to a single species – such as the five fingers on a human hand. There is no reason, other than a quirk of evolution, why we don't have four or even six digits. The eye, for example, has evolved more than forty times in completely unrelated creatures.

But an organ to see with or limbs to move around with are probably both universal features. 'Limbs have evolved independently in different creatures – an octopus has tentacles but they do the same job but with very different structure.' DNA is most likely a parochial feature of Earth, but the idea of evolution of species by natural selection is almost certainly a universal.

What about intelligence? This is not uncommon here on Earth and has also developed in lots of different ways. There are many intelligent creatures on this planet – octopuses, dolphins and whales are all bright. Even mantis shrimps are surprisingly good at solving puzzles if they have to get to their food.

But intelligence would not be enough for us to discover aliens living on another planet. For that we would need an ability for mass cooperation, the capability to develop technology. 'What would give us the possibility of communicating with other planets is not intelligence as such, it's our ability to store anybody's bright ideas in a form that the rest of the culture can access and use,' says Stewart.

Humans today are individually no smarter than previous generations for a good many thousands of years, possibly hundreds of thousands of years. But collectively, our culture can achieve things that were inconceivable a hundred years ago. Extelligence, as Stewart calls it, started with the invention of speech and writing, got going with printing, and is now running riot with the Internet.

Once a species is extelligent, lots of things become possible – they might transcend biology. 'If you're going to talk about intelligent life, for me the important thing is timescales,' says Seth Shostak, a senior astronomer at SETI. Humans went from using radio waves for communication to launching spacecraft in less than seventy years; in one hundred years we might have cracked artificial intelligence. 'Intelligent biology can engineer its own successor and can do it very quickly when you're talking about interstellar communication and interstellar travel. You've already gone beyond the confines of a three-pound brain sitting in a skull on top of your neck.'

While aliens are so out of our experience, guessing motives and intentions if they ever got in touch seems to be, at best, a stab in the dark. The physicist Paul Davies has argued that alien brains, with their different architecture, would no doubt interpret information very differently to ours. What we think of as beautiful or friendly might come across as violent to them, or vice versa.

'Lots of people think that because they would be so wise and knowledgeable they would be peaceful,' says Stewart. 'I don't think you can assume that. I don't think you can put human views onto them and that's a dangerous way of thinking. Aliens are alien. If they exist at all, we cannot assume they're like us.'

What are the chances?

The father of SETI, Frank Drake, reckons that our detection of the first extraterrestrial signal might only be thirty years away, thanks to increasing computing power and the ability to sift more data from more stars ever more quickly. Speaking to *Scientific American* in 2011, he guessed that the number of detectable civilizations in our galaxy right now was about 10,000. It is just a matter of time before we stumble across one.

Future generations of ground-based telescopes will also help in the search, such as the proposed European Extremely Large Telescope (with a 30 m main mirror). This could be operational by 2030, and would be powerful enough to image the atmospheres of faraway planets, looking for chemical signatures that could indicate life.

The SETI Institute is also getting an upgrade, with the continued construction of the Allen Telescope Array. When it has its full complement of 300 dishes scanning the skies, it could examine 1,000 star systems in a couple of years. Shostak is confident that as telescope technology keeps improving, SETI will find an ET signal within the next two decades. 'The bottom line to that is that we will have looked at another million star systems in two dozen years. If this is going to work, it will work soon.'

Is he worried about Hawking's warning about hostile aliens? 'This is an unwarranted fear,' he says. If they're interested in resources, they have ways of finding rocky planets that don't depend on whether we broadcast or not. They could have found us a billion years ago.'

Anyway, if we are genuinely worried about shouting in the stellar jungle, he says, the first thing to do is shut down the BBC, NBC, CBS and all the radars at airports. Those broadcasts have been streaming into space for years, and the oldest is already more than eighty light years from Earth. If we are worried about attracting the attention of passing aliens, we're too late to have stopped them watching every episode of *Big Brother* or *The Jerry Springer Show*. It's hard to guess whether those broadcasts would attract or repel an alien race.

DEATH OF THE SUN

The Sun hangs in the sky, the source of almost all our energy. It lights up our planet, gives us life, allows plants to thrive and makes the world go around. But it will also, one day, blow the Earth apart.

We are in a lucky mid-phase of the Sun's life, a time when its pleasant yellowness is just warm enough to keep our oceans liquid and there is the right amount of radiation to keep plants and animals alive.

In 5 billion years from now, the Sun will emerge from its pleasant middle age and start to get bigger. And hotter. It will eject material, swallow up Mercury and Venus and cause untold damage as it begins its slide towards death. On the Earth, we will have a ringside seat to the destruction, and long after all life has been killed, the remnants of our planet will get involved in the show itself.

The life cycle of stars

Before the Sun, there was a vast cloud of swirling dust and gas. Over millions of years, the hydrogen atoms began to move closer together under the force of gravity and the cloud started to warm up. As it reached 10 million kelvins, the hydrogen began to fuse and the star began to shine. Our Sun began its life fusing hydrogen, and this is how it will spend 90 per cent of its total existence. At present, it is around halfway through the 10 billion years during which a star of its size will normally burn hydrogen.

In around 5 billion years, the Sun will run out of hydrogen to burn, because all of its core will be fused into helium. At this point, the core of the star cannot support itself and it begins to collapse. As it shrinks, it also heats up, and eventually a shell around the shrinking core becomes hot enough to start fusing the hydrogen atoms there. This process will cause the outer layers of the Sun to expand to several tens of times its current diameter. Because the surface at this stage is so far away from the source of the energy, it will cool to around 3,500 kelvins, around half the temperature of the star today, and glow red.

Meanwhile, the helium core continues to contract, and when it reaches around 100 million kelvins, the atoms begin to fuse into carbon and oxygen. The outer surface of the star will become hotter once again, glowing blue and white.

When the helium runs out and the core is entirely composed of carbon and oxygen, the star will once again start to contract and warm up. A shell of helium outside the core starts to fuse and the star once again swells to hundreds of times its current diameter. At this stage it is called a 'symptotic giant branch' star. After around 30 million years, the remnant core will once again start to shrink and heat up.

But a star that starts out the size of the Sun will never get hot enough at its core to start fusing carbon atoms or oxygen atoms there. Instead, the outer layers of the star begin to cool and reach temperatures where electrons can bind themselves to the free-floating atomic nuclei, forming neutral atoms. The outer envelope of the star continues to blow off into space within a relatively sudden 100,000 years, and this halo will contain the raw materials for the formation of planets in future, including carbon, oxygen, neon, sulphur, sodium, argon and chlorine.

These remnants of low-mass stars, called planetary nebulae, are some of the most beautiful objects seen in space, as the hot core of the star lights up the surrounding gas clouds, producing vivid fluorescent colours against the blackness of space. Scientists have been moved to give them names, such as the Cat's Eye, the Starfish Twins, Blue Snowball, Eskimo and the Ant, and they are among the most popular pictures taken by the Hubble space telescope.

The core of the star, meanwhile, will contract for another few thousand years until the electrons become degenerate, which means they cannot be pressed any further together. At this stage, the surface of the remnant has a temperature of around 10,000 kelvins and the object becomes a white dwarf. It contains over half the mass of the Sun packed into a hot (1 million kelvin) ball the size of the Earth. A teaspoon of this object would weigh a tonne on Earth.

Slowly, over billions of years, this white dwarf radiates away energy and cools down further. Finally, when it can shine no longer, all that will be left of the Sun is a lump of ash known as a black dwarf.

What happens to the Earth?

The vast changes in store for the Sun do not bode well for our planet. Even if the Earth manages to survive the first expansion of our star, there is virtually no chance that anything living here will survive. If, somehow, something managed to cling on, it would be faced with a cold, dark, impoverished world without any water. It could not survive long.

'Planetary nebulae are a glimpse into the future of our own solar system,' says Bruce Balick, an astronomer at the University of Washington. 'When the Sun reaches the eleventh hour of its life, it will swell to the size of Earth's present orbit, causing Mercury and Venus to burn up like giant meteors. Earth will escape this fate because the Sun will have blown out some of its material, weakening its gravity so that our planet slips into a new, larger orbit.'

The burned-out husks of Mercury and Venus will find themselves orbiting the Sun from inside the red giant. On Earth, if we have survived, sunrises and sunsets will take hours, and at noon, the gigantic red Sun will take up half the sky. The oceans, along with the atmosphere, will boil away into space. 'The view won't be very different than that within a kiln,' says Balick. 'The intense radiant heat will transform the surface to a thick layer of pottery.' In all, a biblical view of hell.

Balick does however see one up side to the rather compromised situation: the Earth would be able to witness from the inside the formation, millions of years after the red giant, of a planetary nebula. 'The Sun will eject its outer layers in an extreme version of the present-day solar wind,' he says. 'Eventually the red behemoth will be stripped to its core, which will quickly settle down as a white dwarf star. Lit by this blue-hot pinprick, objects on Earth will cast sharp-edged, pitch-black shadows; sunrise and sunset will take no longer than an eyeblink. Exposed rock will turn to plasma as ultraviolet radiation from the dwarf destroys all molecular bonds, coating the surface with an eerie iridescent fog, constantly lifting and swirling. As the dwarf radiates away its energy, it will fade into a cold, dark cinder. Thus, our world will end first in fire, then in ice.'

The white dwarf would rise in the Earth's sky as an intense spot of light, one hundred times brighter than the Sun is today but no bigger than Venus. The light from this star would bake the Earth's rocks and rip apart the molecules on the surface, creating a new atmosphere of free electrons.

What can we do?

It seems there is little we humans can do in the face of such stellar inevitability. In 2001, however, scientists at the University of Michigan came up with an idea for a global-scale project that could shift the Earth's orbit by small amounts by engineering close passes to asteroids and comets. The gravitational attraction of these bodies would pull the Earth into an incrementally higher orbit around the Sun. Over billions of years, this might be far enough outside today's orbit to save us from the worst ravages of the Sun's red giant phase. It's a dangerous idea, though, because it would take thousands if not millions of close passes with big rocks in space to move our orbit. And engineers would have to make sure that none of these rocks ever hit the Earth, or else they would cause a different type of doomsday event (see p.143). If it somehow worked, it would only save the Earth for a short while, until the Sun eventually dimmed into nothingness.

Our best option, then, is to leave the Earth and colonize a planet far from here. We certainly have enough time to invent, test and build the required technology. There is also enough time to identify and reach the closest planets at sub-light speeds, though each trip might take several generations. It sounds far-fetched, perhaps, but it is one of the inevitable things about nature that stars, and therefore their planets, have a finite lifetime. Like it or not, we cannot stay on the Earth forever.

GALACTIC COLLISION

There are some parts of our cosmic ballet that sound far too big to think about, things we just have to watch while resigning ourselves to their effects. The Sun expanding and engulfing the Earth is one event that will mark a turning point for our species. Another phenomenon that will happen at around the same time sounds on the surface far more violent: the wholesale collision of our galaxy with another.

Look at the sky, and virtually all of the galaxies we can see are speeding away from us, an echo of the Big Bang that reverberates to this day in the expansion of space and the universe. All of them, that is, except the galaxy M31, which is moving towards us. At its current speed of 120 km/s (75 miles per second), M31 (also known as Andromeda) will reach the Milky Way some time in the next 5 billion years. If we have managed to live that long, this cosmic smash might well finish us off.

Our galactic neighbourhood

The Milky Way is part of a loose collection of a few dozen galaxies, called the Local Group. From a human scale, calling it a collection might seem odd, given that the galaxies themselves are separated by up to one hundred times their own diameters. But, bound and governed by gravity, these cosmic objects are jostled around their vast backdrop. The galaxies move slowly into and out of each other's paths, and over the full lifetime of a typical galaxy – estimates range from 10 billion to 20 billion years – a collision is inevitable.

In our Local Group, the Milky Way and Andromeda are the biggest characters, roughly the same size as each other at around 270 billion times the mass of the Sun. Andromeda is the smaller cousin, a swirling mass of several billion stars that is currently 2.5 million light years away.

Until twenty years ago, astronomers had doubts about whether galaxies would ever collide, given the huge spaces between them. But a combination of improved observation and better computer models that can simulate galactic motions has revealed that collisions are not only possible, they are probably more common than anyone thought. According to astronomers Joshua Barnes, Lars Hernquist and Francois Schweizer, the available evidence suggests that colliding galaxies often merge into a new kind of object. 'We have become increasingly convinced that such collisions control the evolution of many galaxies and lead to the formation of a variety of peculiar objects, possibly including the distant and extraordinarily luminous quasars.'

How close are we to a collision? What would happen?

Even in the most violent encounter, it is unlikely that the stars themselves will actually hit each other, given the vast spaces between them. In many large galactic clusters, pairs of galaxies are thought to come towards each other at high speeds – thousands of kilometres per second – and pass right through each other, apparently causing no damage in the process. 'Oddly enough, if the same galaxies were to approach at a velocity of only a few hundred kilometers per second, they would violently disrupt one another and probably would merge within a few hundred million years,' say Barnes, Hernquist and Schweizer in an article for *Scientific American*. 'Such seemingly paradoxical behavior reflects the fact that galactic interactions are governed by gravitational forces. The slower an encounter between two galaxies, the more time there is for gravity to produce huge, disruptive tides and the greater the resulting damage.'

But that does not mean the components of the galaxies feel nothing in the process. Just as the Earth's oceans feel the attraction of the Moon in the creation of the tides, so the passing galaxies will create a gravitational tide as they pass each other.

'Tides between galaxies are much more disruptive than terrestrial ocean tides because galaxies pass much closer to each other, relative to their size, than do the earth and moon,' say Barnes, Hernquist and Schweizer. 'If the moon orbited the earth at half its present distance, the gravitational force it exerts would increase by a factor of four, because gravity is inversely proportional to the square of the distance. But the difference between the forces on the near and far sides of the earth, which is what determines the height of the tides, would increase by a factor of eight. In other words, tidal forces are inversely proportional to the cube of the distance. In close collisions the tidal forces between a pair of galaxies can be strong enough to rip both apart.'

The collision between Andromeda and the Milky Way is

likely to be a multi-part process. Before the full merger in around 5 billion years, the galaxies will pass and hit each other with glancing blows. This will happen first in around 2 billion years, and there is a slight chance that our solar system will be thrown out of the Milky Way in the process, part of a tidal stream of material dragged out of our galaxy by Andromeda.

It will only be a few billion years before Andromeda is back to repeat the process. It will take a few swipes and several more billion years before the two galaxies settle down into a merged mass of stars that orbit a common centre of gravity. The combined galaxy itself would be a huge elliptical object, without the sweeping arms we know of our Milky Way.

The colossal energies released every time the galaxies meet would cause shock waves to pass through the vast clouds of dust and gas that float between the stars. Each collision would compress the hydrogen in the clouds sufficiently to start the process of fusion, giving birth to a flurry of new stars every time our galaxies get close.

The timescales over which the collisions will occur are unimaginably vast. By the time the situation has settled down, in more than 7 billion years' time, the Earth will be having problems of its own as the Sun swells up and swallows its closest planets. Assuming we are somehow still alive as a species to watch the formation of the new mega-galaxy, however, what will be the fate of our solar system?

A computer simulation by scientists at the Harvard-Smithsonian Center for Astrophysics reckons that our Sun will keep its remaining planets, but there is a high probability that the ensemble will be pushed away from the centre of the newly merged galaxy – perhaps up to 100,000 light years from the twin black holes (one coming from the centre of each original galaxy) compared to our distance of 26,000 light years from the centre of the Milky Way at present.

What can we do? How likely is the collision?

It seems vaguely ridiculous to think about whether we could do anything about an event of this magnitude that is predicted to happen so far in the future. But for anyone harbouring hopes, it is worth pointing out that astronomers are not yet certain that the collisions will definitely happen.

Andromeda and the Milky Way are moving towards each other, of that there is no doubt. They will also come very close to each other and their mutual gravitational attraction will have tide effects on the objects within the galaxies as they pass. But no one is quite sure how the two galaxies will come together.

The uncertainty arises from the fact that astronomers know how fast Andromeda is spinning, but cannot measure its transverse velocity. This means we do not know if Andromeda will hit us square on, or whether it will be a glancing blow where only the outskirts of the galaxies (or just the haloes of dark matter that surround them) make contact.

Either way, let us hope that our descendants are watching the whole thing from a vantage point very far away.

THE END OF TIME

Good old time. Sitting in the backdrop of our daily lives, giving them order, providing the arena upon which all of us can get things done. It will always be there, right? Wrong. Some theories suggest that, at some point in the future, there is no 'after'.

It might have suffered at the hands of Albert Einstein, who was convinced that time was part of a bigger, more profound arena called space–time, but to the rest of us it stays the same. Ticking clocks, shifting seasons, babies growing up. We and our world might finish up as dust, but the universe will carry on; nothing really stops.

But that is not enough for physicists. In the past century, scientists have pondered on the basics of what time is and how it really fits into the scientific picture of the universe. Their ideas are startling and dangerous: perhaps time does not really exist, say some; perhaps it is draining away from the universe, argue others. Unfortunately for the survival of our universe, the death of time is frighteningly possible.

What is time?

Trying to define time feels something like asking what air is: it is a property of the world that is just there; it exists and without it our lives would, well, stop. Physicists and philosophers, though, always want to go further. Is time simply a way of labelling a sequence of events so that we know what has happened, in what order and how far apart? Or, like space, are we actually measuring some sort of 'stuff' when we measure seconds and hours? We might not know what it is, but this 'stuff' ticks away in the background.

Our experience of time is that it flows from one moment to the next. We are in a present that keeps moving, leaving a set of events in the past in our memories. 'We have a deep intuition that the future is open until it becomes present and that the past is fixed. As time flows, this structure of fixed past, immediate present and open future gets carried forward in time. This structure is built into our language, thought and behavior. How we live our lives hangs on it,' says Craig Callender, a philosopher at the University of California, San Diego.

But this natural way of thinking is not reflected in science. 'The equations of physics do not tell us which events are occurring right now – they are like a map without the "you are here" symbol,' says Callender. 'The present moment does not exist in them, and therefore neither does the flow of time. Additionally, Albert Einstein's theories of relativity suggest not only that there is no single special present but also that all moments are equally real. Fundamentally, the future is no more open than the past.'

In Isaac Newton's day, it seemed as though the universe

came with a clock that split our experience of the world into segments that we call seconds, minutes or hours. During these moments, things would happen. Since the middle of the 19th century, however, scientists have known that scientific laws do not require these segments of time to proceed in any particular direction. In other words, there is nothing inherent in physics that says that time must move 'forward' rather than 'backward'.

The laws of physics work perfectly whether time is moving forwards or backwards, and the 19th-century Austrian physicist Ludwig Boltzmann went so far as to suggest that the difference between the past and the future was something not intrinsic to time itself but simply a result of differences in how matter in the universe was arranged.

Albert Einstein's theories of special and general relativity in the early 20th century put more nails into the coffin of the Newtonian 'universal clock' idea. Special relativity made time part of the coordinate system with space, melding them into a four-dimensional space–time where different people moving at different speeds would feel seconds ticking at different rates. In general relativity, gravity itself distorts the passage of time, so a second next to a massive star 'ticks' at a different rate to a second in deep space, where there is less gravity.

These developments have left us with a conundrum. 'You cannot generally think of the world as unfolding, tick by tick, according to a single time parameter,' says Callender. 'In extreme situations, the world might not be carvable into instants of time at all. It then becomes impossible to say that an event happened before or after another.'

Which begs the question that many physicists have been pondering for decades – does time exist anyway? 'The idea of a timeless reality is initially so startling that it is hard to see how it could be coherent,' says Callender. 'Everything we do, we do in time. The world is a series of events strung together by time. Anyone can see that my hair is graying, that objects move, and so on. We see change, and change is the variation of properties with respect to time. Without time, the world would be completely still. A timeless theory faces the challenge of explaining how we see change if the world is not really changing.'

Where time ends

General relativity, Einstein's theory of how gravity works, predicts the existence of something else disturbing to anyone still stuck in the Newtonian universe of time: singularities. These are points of infinite density in space where matter is squashed, physics essentially breaks down and time stops. They are the objects at the centre of black holes, the result of supermassive stars that have collapsed into points smaller than the following full stop. The intense gravity sucks in anything that gets too close (within a boundary called the event horizon), and at the singularity itself, time does not move at all. Get into one of these, and there is no such thing as 'after'.

There are other places where time seems to disappear. Much of the universe is known to be missing – the matter we are made of accounts for only 4 per cent of the mass in the universe, with the rest split 20–76 between dark matter and dark energy. This latter substance (no one really knows what it is) seems to be pushing the galaxies apart, keeping the universe expanding despite the gravitational attraction between all the familiar and dark matter.

In 2007, José Senovilla, Marc Mars and Raül Vera of the

University of the Basque Country, Bilbao, and the University of Salamanca, Spain, proposed an alternative explanation for the effects of dark energy on the universe. In a paper for *Physical Review D*, they suggested that, rather then there being an anti-gravity force at work in the universe, the effects we observe are due to time slowing down as it leaks away. We do not notice it at an everyday level, but it should become apparent in the motion of galaxies as they move over the span of billions of years.

Our observed expansion of the universe, therefore, is an illusion. In fact, time is slowing down.

The idea is based around a concept in string theory, which is an attempt to explain the universe at a more fundamental level than current scientific ideas. In string theory, all particles are made of multi-dimensional strands of energy, each vibrating at a different frequency. One of the resulting ideas is that our universe exists on a membrane (known as a 'brane'), part of a 'bulk' of other branes and universes. 'Ordinarily, we are free to roam around our 4-D prison,' says writer George Musser, explaining Senovilla's idea in an article for *Scientific American*. 'But if the brane is blown around fiercely enough, all we can do is hold on for dear life; we can no longer move. Specifically, we would have to go faster than the speed of light to make any headway moving along the brane, and we cannot do that. All processes involve some type of movement, so they all grind to a halt.'

Anyone on the brane would be unaware that this is happening to them because their clocks would stop too – they would have no way of telling that time was turning into space. 'All we would see is that objects such as galaxies seemed to be speeding up,' says Musser. 'Eerily, that is exactly what astronomers really do see and usually attribute to some unknown kind of "dark energy". Could the acceleration instead be the swan song of time?'

Is it likely?

Whatever time is, we know that it was created 13.7 billion years ago with the Big Bang, and if the universe were to end, whatever time is would go with it. When pondering the fate of the universe, cosmologists think the most likely result is that everything will continue expanding for ever, eventually whimpering to an end when energy is so sparsely spread that, though time technically never ends, it has become meaningless. In this 'heat death' scenario, everything is in equilibrium and any process or interaction is quickly undone by a process that goes in the opposite direction.

'Physicists argue it both ways,' says Musser. 'Some think time does end. The trouble with this option is that the known laws of physics operate within time and describe how things move and evolve. Time's end points are off the reservation; they would have to be governed not just by a new law of physics but by a new type of law of physics, one that eschews temporal concepts such as motion and change in favor of timeless ones such as geometric elegance.'

Even if the universe continued on forever, without time, we would not live.

STRANGELETS

The subatomic world has plenty of surprises, though typically none of them are cataclysmic. There is a secret, though, in the laws of physics. A particle so deadly that its mere presence would spell the end for our planet.

Perhaps predictably, the weirdest stuff comes from one of the most successful, but also most bizarre, physical theories ever devised: quantum mechanics. This science of the very small is filled with peculiar ideas, such as the inability to predict exactly where subatomic particles are or what they are doing. Some interpretations even suggest the existence of multiple worlds, each new one branching off every time we make a decision. But none of these ideas, however odd, is necessarily cataclysmic.

Yet the equations harbour a terrifying secret. Among the descriptions of fundamental particles and the ways they can bind together to form the everyday matter we are familiar with, something ghastly is hiding – a theoretical particle so stable that it can transform any other particle of matter into a copy of itself.

There would be no energetic coercion going on. The laws of physics state baldly that if this particle, the strangelet, came into contact with a particle of normal matter (made of protons and neutrons), the latter would somehow recognize that it was in a hopelessly inefficient energy state and immediately reorganize itself into a strangelet too. These copies would then go on to convert other particles into still more strangelets.

It is the ultimate doomsday weapon: in just a few short hours, a small chunk of these particles could turn an entire planet into a uniform, featureless mass. Everything that planet was made of, everything that was on it, would be no more.

What is a strangelet?

To understand what a strangelet is, we need to step back into the basics of what makes up the stuff of the universe.

The standard model of particle physics gives a precise quantum mechanical description of all the subatomic particles that are known to exist. It says that all matter particles consist of a combination of six quarks (some of which make up protons and neutrons, others that are so heavy they only survive for fractions of seconds before decaying into lighter particles) and six leptons (including the electron and neutrinos). There are also particles (called bosons) that carry the fundamental forces, which include the particle of light, the photon and gluons that stick quarks together in the nucleus of an atom.

The strangelet contains one of the lesser-seen quarks. There are three 'generations' of quarks, each one featuring particles that are more massive than those in the set before it: up and down, strange and charm, top and bottom. Only two of these quarks – up and down – affect us in our daily lives. A proton consists of two up quarks and a down quark; a neutron is made of two down quarks and a single up quark.

STRANGELETS

A strangelet is a hypothetical particle consisting of equal numbers of up, down and strange quarks. Because strange quarks are so heavy, this composite would be the same size as a small atomic nucleus that might otherwise contain scores of other up and down quarks. In normal life, a strange quark is unstable and decays very soon after it is formed into lighter quarks.

However, it has been hypothesised that if lots of up, down and strange quarks got together, the resulting mass would be less prone to decay. According to the strange-matter hypothesis, thought up by, among others, Ed Witten at the Institute of Advanced Study in Princeton, a strangelet with lots of quarks would be even more stable than a normal atomic nucleus.

If this strangelet were to collide with a normal nucleus, the conversion of the latter into a strangelet would take a billionth of a second and release energy that would then be available for other nucleus conversions. One by one, every atomic nucleus in a collection of ordinary matter – the Earth, say – would be transformed into strangelets, leaving our planet a hot lump of strange matter.

One science-fiction writer who did think up something similar was the visionary Kurt Vonnegut. In *Cat's Cradle*, he describes a fictional material called Ice Nine, which is supposed to be a more stable form of water that melts at 45.8°C instead of 0°C. When Ice Nine comes into contact with normal water, it acts as a catalyst to solidify the entire body of water. Inevitably, somebody uses this material to solidify all of the Earth's oceans.

Do strangelets exist?

There is nothing to prove that vast clouds of strangelets are not floating undetected in deep space today. They might be produced naturally in cosmic collisions, and could be part of the explanation for the mysterious dark matter that scientists know makes up a quarter of the mass of the universe, but which we cannot see. If it is there, there is little we can do about it drifting into our solar system.

More worrying is the possibility of creating a strangelet on Earth. Whether we are at the mercy of death by strangelet has been considered seriously several times by physicists, usually while trying to calm public fears over the building of ever-larger particle accelerators that might accidentally produce strangelets in their high-energy collisions.

Before scientists fired up the Relativistic Heavy Ion Collider (RHIC) in 2000 in the United States, they carried out a study into the various catastrophic events that might occur by accident when particles are smashed at such high energies.

Describing the RHIC in 1999, Sheldon Glashow and Richard Wilson, physicists at Harvard University, said: 'Beams of highly charged gold or lead atoms (the heavy ions) travelling at relativistic speeds (99.95% of light speed) will speed in opposite directions around circular racetracks before colliding. RHIC is truly an atom smasher: nucleus–nucleus impacts, taking place thousands of times per second, will each produce thousands of secondary particles. These incredibly complex "events" will be recorded by sophisticated detectors and analysed at supercomputers or farmed out to a world consortium of smaller computers. RHIC will study matter at densities and temperatures never seen in the laboratory. On a small scale, it will reproduce the extreme conditions that reigned in the early universe, conditions under which the constituents of ordinary matter are expected to be liberated as a quark–gluon plasma.'

The scientists go on to describe the conclusions of two safety studies. 'Both groups include theorists who were among the first to speculate that lumps of strange matter called strangelets – which contain many strange quarks as well as the usual up and down quarks that make up atomic nuclei – might be more stable than ordinary matter. If strangelets exist (which is conceivable), and if they form reasonably stable lumps (which is unlikely), and if they are negatively charged (although the theory strongly favours positive charges), and if tiny strangelets can be created at RHIC (which is exceedingly unlikely), then there just might be a problem. A newborn strangelet could engulf atomic nuclei, growing relentlessly and ultimately consuming the Earth. The word "unlikely", however many times it is repeated, just isn't enough to assuage our fears of this total disaster.'

Is it likely?

Glashow points to experiments in nature to allay fears that strangelets can be created easily. Cosmic rays (mostly very energetic protons in space) stream around the universe at speeds near to that of light; anything of appreciable size in the cosmos will be buffeted by these particles all the time. Glashow uses the Earth's Moon as a good example of a natural experiment.

'Lacking a protective atmosphere, with a surface rich in mid-sized atoms such as iron, it is a plausible target on which incident cosmic rays of iron (or larger) nuclei with RHIC energies could produce strangelets,' he says. 'Yet countless collisions over billions of years have left the Moon intact.'

Before you get too comfortable, though, here's one more study to think about. The risk of a doomsday scenario in which high-energy physics experiments trigger the destruction of the Earth might have been estimated to be tiny, but this may give a false sense of security, according to Max Tegmark and Nick Bostrom, respectively a physicist at the Massachusetts Institute of Technology and a philosopher at the Future of Humanity Institute at the University of Oxford. 'The fact that the Earth has survived for so long does not necessarily mean that such disasters are unlikely, because observers are, by definition, in places that have avoided destruction,' they pointed out in *Nature* in 2005.

Given that life on Earth has survived for nearly 4 billion years, it might be easy to assume that natural catastrophic events are extremely rare. Unfortunately, say the researchers, this argument is flawed, because it fails to take into account 'an observation-selection effect, whereby observers are precluded from noting anything other than that their own species has survived up to the point when the observation is made. If it takes at least 4.6 [billion years] for intelligent observers to arise, then the mere observation that Earth has survived for this duration cannot even give us grounds for rejecting with 99% confidence the hypothesis that the average cosmic neighbourhood is typically sterilized, say, every 1,000 years. The observation-selection effect guarantees that we would find ourselves in a lucky situation, no matter how frequent the sterilization events.'

Using information on planet-formation rates, the distribution of birth dates for intelligent species, they add, can be calculated under different assumptions about the rate of cosmic sterilization. 'Combining this with information about our own temporal location enables us to conclude that the cosmic sterilization rate for a habitable planet is, at most, of the order of 1 per 1.1 [billion years] at 99.9% confidence.'

So no need to get too concerned yet.

GENETICS

GENETIC SUPERHUMANS

With the first draft sequence of the human genome in 2000, scientists finally gathered the tools that guided the biochemistry of life. Their first task has been to understand and manipulate that biochemistry to banish disease. But what about using that knowledge to make humans even better, faster, stronger, more intelligent?

If genetic technology could be used to stop disease, why not select genes to improve a person's physical or mental ability? If someone has the money and the inclination, might they programme their children to be superachievers? If the technology is available, why leave human evolution to the slow and random process of natural selection, when instead you can ramp up the benefits and the speed with active design of a genome?

Successive generations of enhanced children might even start to separate into a new species, different from the run-of-the-mill humans we see today, with their ragbag mixture of advantages and flaws. These enhanced people will be all-perfect, and ultimately, might even outcompete their unenhanced cousins into extinction. Thanks to genetic technology, we could create a new species of humans, and perhaps bring an end to *Homo sapiens*.

What is possible with genetic modification?

In 1953, Francis Crick and James Watson sent molecular biology into a spin by detailing the molecular structure of DNA, the molecule that encodes the instructions to make a living organism. This double-helix molecule, they said, was made of four nucleotide molecules (adenine, thymine, guanine and cytosine) in a specific sequence. They were bound in pairs along the length of the DNA – adenine with thymine, guanine with cytosine – and held in place by a backbone of phosphate and sugar molecules.

Almost fifty years later, President Bill Clinton and UK Prime Minister Tony Blair held a joint press conference to announce that scientists had finished sequencing the 3 billion nucleotide letters contained in a human genome.

It was a huge moment. Understanding the DNA contained in a person is the key to understanding most, if not all, human diseases, including cancers and untreatable neurodegenerative conditions such as Alzheimer's. DNA is also an important factor in the range of normal physical characteristics such as height, cognitive ability, muscle mass and rate of metabolism. The Human Genome Project revealed the basic letters and vocabulary with which nature writes its grand narratives of life. For the past ten years scientists have been trying to work out the stories that our genes are telling us. Curing disease is at one end of the applications possible with this genetic knowledge. Enhancing healthy people is at the other.

How easy is genetic modification?

The idea that you might be able to reprogramme your child's genome to make sure it has the best possible traits has been the cornerstone of popular hopes for genetic technology. 'When will it possible to remove harmful genetic mutations from my future baby?' you might ask. And while you're doing that, why not give him blue eyes and an increased number of fast-twitch muscle fibres so that he can grow up to be a star sprinter?

For now, genetic reprogramming is hard work and the results are not guaranteed. There are many technical reasons: selecting a gene means that one or other of the parents has to have that gene to start with. And if you want a baby with a specific gene, you would need to create several dozen embryos at once and somehow select those with the most suitable genes. This technique, called pre-implantation genetic diagnosis (PGD), is used in combination with in-vitro fertilization (IVF).

If you want to programme a whole genome with specific genes, however, PGD is very clunky. Altering a lot of specific genes at once would require the creation of scores (perhaps hundreds) of embryos, and then sorting through them to find a perfect match. Think of the number of eggs that would be wasted if you wanted to pick lots of different genes for your baby; and bear in mind that even if you found a perfect set, the chances of an embryo growing to full term in an IVF cycle are not guaranteed.

In addition, our understanding of genes is nowhere near sophisticated enough to be so prescriptive about the kinds of results a genetic programmer might demand. Diseases caused by a single gene defect are rare – a classic example is cystic fibrosis, a hereditary condition that can cause the airways in the lungs to become clogged with sticky mucus. Scientists have pinpointed that it is caused when both copies of the CFTR gene do not work properly. Most diseases have hundreds of genes involved at varying levels. And for physical characteristics, genes are only part of the picture, with the rest down to environmental factors. Having the genes for sprinting muscles is only useful if you train sufficiently to make them work.

Another way of reprogramming a genome is to alter the germ-line cell (the egg or sperm) of a parent. Scientists can already stop the action of particular genes in animals for the purposes of research, and this is done by modifying germ-line cells. The technique is full of dangers, though. Around 15 per cent of such experiments tried out on mice have proved lethal, and an even higher number produce disabilities in the animals. Many genes have multiple uses in humans – the same gene is associated with a boost in IQ, for example, but also a muscle condition that can leave sufferers in a wheelchair.

A more promising way of altering gene expression is a technique that does not involve germ-line cells or programming foetuses. Instead it uses the body's natural mechanism for deciding which genes are active in which cells and to what degree. This method, called epigenetics, is defined as the changes in gene expression that have nothing to do with the sequence of DNA. Epigenetics can change a huge range of things in an organism, from the shape of flowers to the colour of a fruit fly's eyes.

One procedure involves attaching molecules, called methyl groups, on to a specific part of a gene. In effect, the methyl group silences the action of the gene. The methyl group can be a temporary addition, and can even be introduced via local environmental factors, including chemicals or food.

The problem with unchecked modification

One day, we will overcome the technology hurdles. Imagine a future where you are able to choose what your kids look like (based, of course, on your own looks), whether they are tall or short, how good their eyesight is, whether they will be sprinters or marathon runners, how clever they will be and whether they will tend towards kindness or being selfish. Of course you will already have excluded any genes that might cause disease.

If these technologies become available, it is hard to imagine that people will not make use of them. The world is a competitive place; any advantage you can give your child has to be worth it, right? 'This has been a focus for many opponents of germ-line genetic engineering who worry that it will widen the gap between haves and have-nots,' says Nick Bostrom, philosopher at the University of Oxford and director of its Future of Humanity Institute. 'Today, children from wealthy homes enjoy many environmental privileges, including access to better schools and social networks. Arguably, this constitutes an inequity against children from poor homes. We can imagine scenarios where such inequities grow much larger thanks to genetic interventions that only the rich can afford, adding genetic advantages to the environmental advantages already benefiting privileged children. We could even speculate about the members of the privileged stratum of society eventually enhancing themselves and their offspring to a point where the human species, for many practical purposes, splits into two or more species that have little in common except a shared evolutionary history.'

These genetically privileged people, says Bostrom, might become ageless, healthy supergeniuses of flawless physical beauty, who are graced with a sparkling wit and a disarmingly self-deprecating sense of humour, radiating warmth, empathetic charm and relaxed confidence. 'Everyone else would remain as people are today but perhaps deprived of some of their self-respect and suffering occasional bouts of envy. The mobility between the lower and the upper classes might disappear, and a child born to poor parents, lacking genetic enhancements, might find it impossible to successfully compete against the super-children of the rich. Even if no discrimination or exploitation of the lower class occurred, there is still something disturbing about the prospect of a society with such extreme inequalities.'

Might there be a scenario in which there is so much tension between the enhanced and the unenhanced that they eventually go to war? Would the enhanced humans, with their superior strength and intellect, simply enslave their normal-human cousins? As far as modern humans go, it would spell the end.

Is it likely?

Our understanding of genes will continue to get better, and scientists will improve their ability to genetically modify humans. You could argue about how much will be possible, but there is little use in betting against what is essentially a technology issue. The question about its impacts, and the potential doomsday consequences for the modern human race, comes down to how the technology is made available, implemented and regulated.

One way of preventing potential problems is to ban everything. This might work among the law-abiding, but if the fruits of genetic enhancement were potentially big, a black market would soon develop and the two-tier society would still evolve. Bostrom suggests the opposite as a way to stem the problems. To counteract some of the inequality-increasing tendencies of enhancement technology, governments could widen access to the technology by subsidizing it or providing it for free to children of poor parents. 'In cases where the enhancement has considerable positive externalities, such a policy may actually benefit everybody, not just the recipients of the subsidy,' he says. 'In other cases, we could support the policy on the basis of social justice and solidarity.'

Which might sound like an overly rosy view of the world. But for anyone more cynical, Bostrom has another example of why genetic enhancements might not be the danger that pessimists imagine, in particular why a war between the enhanced and unenhanced is highly unlikely. Right now, the tallest 90 per cent of the population could, in principle, band together and kill or enslave the remaining, shorter 10 per cent of the human race, he says. 'That this does not happen suggests that a well-organized society can hold together even if it contains many possible coalitions of people sharing some attribute such that, if they unified under one banner, would make them capable of exterminating the rest.'

It might be possible to use genetic technology to design a new human species and eventually destroy the original, unmodified humans. The only way to prevent that happening is to hope that humanity itself is not designed out in our quest for perfection.

DYSGENICS

Is our modern lifestyle interfering with natural selection? With better medicine and healthcare, are we allowing the spread of 'undesirable' DNA, the stuff that would once have been weeded out of the human gene pool by evolution? Could this lead to big problems for our species?

A prosperous, happy society will need lots of intelligent people who treat each other with respect. They will look out for each other and come up with ever-better ways to improve their collective lot. Genes that contribute to health and intelligence are good things, and if natural selection was left to get on with it, these would be prevalent in society. Any 'bad' genes for disease, disability or mental deficiency should be weeded out by the principles outlined by Charles Darwin in the 19th century. But medical technology has put a stop to this natural mechanism.

Even worse, the genes that contribute to the more preferable traits in society might themselves be at risk. If IQ is determined, to some extent, by genes, and people with high IQs do not have as many babies as those with lower IQ points, intelligence across the population will suffer. If being good to others is partly genetic, and good people are similarly less fecund than their bad cousins, then 'badness' genes might be growing more popular.

'Currently it seems that there is a negative correlation in some places between intellectual achievement and fertility,' says Nick Bostrom, a philosopher and the head of the Future of Humanity Institute at the University of Oxford. 'If such selection were to operate over a long period of time, we might evolve into a less brainy but more fertile species, *Homo philoprogenitus* ("lover of many offspring").'

Might 'good' genes eventually be swamped by 'bad' genes among humans?

We are all mutants

Steve Jones, a geneticist at University College London, thinks that our ability to cure diseases that would previously have killed us, the ways we move around the world, and our huge level of control over our bodies and our environments have all conspired to take power away from the natural forces of evolution.

Variations among humans are driven by genetic mutation. Every time a cell divides, the DNA within can mutate. Most changes have no overall effect and are never passed on to children. But rarely, the mutations can change the way some part of the body looks or works, and even more rarely, that change is fatal. As we age, the number of mutations adds up.

The process by which a particular mutation becomes more common in a population is called selection. Within the past 5,000 years, for example, a gene for skin colour mutated to give someone white as opposed to dark skin. The white skin, which has the advantage of being better at making vitamin D from sunlight, became useful for those living in northern Europe, where there was less sunlight. Selection pressure means that the white gene variant appears in 99 per cent of Europeans, whereas 99 per cent of Africans maintain the dark skin variant.

Mutation and selection are the raw materials for evolution, the stuff that the random forces of nature can shape into species. But where once a person's genes had complete sway over their longevity, leaving only those with the 'best' genes to survive and pass on their DNA, modern medicine and lifestyles have levelled the playing field, diminishing the stuff that evolution can play with. A lack of vitamin D does not need any genetic fixes today; it can be easily treated with food supplements. The genetic mutations we collect as we age might no longer kill us as we learn more about how to fix them. Where people in one part of the world might have been wiped out by a particular infection, medicine or a re-engineering of the local environment can save lives.

More important, given that previously dangerous genetic mutations can now routinely survive and be passed on to children, Jones questions what power is left for natural selection in humans.

The dysgenic dystopia

Dysgenics, the idea that bad genes are spreading in our population, was a term coined in the 1970s by the American physicist and Nobel laureate William Shockley. In the previous decades, he had developed a theory about a possible connection between race and intelligence. His argument, which used data from IQ tests on American soldiers, suggested that blacks were genetically inferior to whites and that intelligence in people who were mixed race depended on how much 'white blood' they had in their ancestry. As a result, he called on people with low IQs to undergo voluntary sterilization, to stop them passing on their genes to a new generation of children.

The ideas are certainly dubious by today's standards. But Shockley was not alone in thinking that undesirable genes could spread and compromise the progress of the human race – even the great naturalist Charles Darwin was not immune to pessimism about the fate of humans, according to Alfred Russel Wallace, the biologist and co-founder of the theory of natural selection. 'In one of my last conversations with Darwin he expressed himself very gloomily on the future of humanity, on the ground that in our modern civilisation natural selection had no play and the fittest did not survive,' wrote Wallace. 'It is notorious that our population is more largely renewed in each generation from the lower than from the middle and upper classes.' The implication, of course, is that those in the lower classes were somehow biologically inferior, and that their children would be inferior too.

The early 20th century was marked by several attempts by scientists and governments to justify eugenics, a movement founded by Darwin's half-cousin Francis Galton, and which tried to 'improve' the human race by breeding out deleterious inherited traits, in everything from intelligence to temperament and morals.

'British eugenicists tended to look for ways to perpetuate the privileged classes, whereas American eugenicists tried to slow down the supposedly corrupting influence of the degenerate classes,' wrote biologists David Micklos and Elof Carlson in an article on the history of eugenics, published in *Nature*

Reviews Genetics. 'Eugenics researchers attempted to trace the inheritance of dysgenic traits through a family tree, and constructed pedigrees by interviewing living family members and by scrutinizing the records of poorhouses, prisons and insane asylums.'

In a precursor to some of Shockley's ideas, the French aristocrat Joseph Arthur Gobineau proposed in the mid-19th century that the mixing of races was at the root of genetic deterioration. And eugenicists sometimes tried to provide a supposed scientific rationale for existing racial prejudice – in his influential book *The Passing of the Great Race*, Madison Grant warned that racial mixing was a social crime that would lead to the demise of white civilization.

For American eugenicists, 'feeble-mindedness' became an important idea around 1910. It was linked to abnormal behaviour, promiscuity, criminality and social dependency. The influence of the eugenics lobby can be seen in the first sterilization law passed in Indiana in 1907, on the advice of prison physician Harry Clay Sharp. 'Speaking at meetings of the American Medical Association, Sharp convinced fellow physicians to lobby their legislatures for laws to allow the involuntary sterilization of sex offenders, habitual criminals, epileptics, the "feebleminded" and "hereditary defectives", say Micklos and Carlson. 'The intent was to prevent alleged degenerates from breeding with each other or from contaminating "good genetic stock" by reproducing with normal people.'

Marian Van Court, who runs an organization called Future Generations that tries to rehabilitate the image of eugenics, explains the potential dysgenic problem in terms of intelligence. 'Since IQ is positively correlated to a number of desirable traits (such as altruism, anti-authoritarian attitudes, and middle-class values of hard work, thrift, and sacrifice), when IQ declines, so do these traits,' she says. 'People with low IQs are far more likely to become criminals, so the fact that our genetic potential for intelligence is declining means our genetic potential for crime is increasing. Moreover, some evidence suggests that despite lengthy sojourns in jail, criminals still manage to procreate at a faster rate than the rest of us.'

Van Court cites research by Richard Lynn of the University of Ulster to underline her point. He found that criminals in London had nearly twice as many offspring as non-criminals. 'In demographic studies of fertility, the entire category of underclass males is frequently omitted because reliable data on their offspring simply can't be obtained – their sexual behavior is often promiscuous, and their relationships transient,' she says. 'Since twin studies and adoption studies have established that there is a substantial genetic component to criminality, the higher fertility of criminals significantly increases the genetic potential for criminality in the population.'

If you believe this description of society, criminality is on the rise and intelligence and good temperament are on the wane, at least genetically speaking.

Purposely choosing 'undesirable' genes

Eugenics is not the only thing to note if you are thinking about the spread of so-called 'undesirable' genes. Most people with normal hearing might consider deafness to be a terrible affliction that should be avoided or cured if possible. But many of those born deaf think of themselves as a cultural minority with shared customs and ways of living that should be preserved.

'An analogy is drawn with members of minority ethnic communities, who share a common language,' says Tom Shakespeare, a bioethicist at the University of Newcastle. 'Deaf people (who adopt a capital letter to signify this distinctiveness) share a culture based around sign language. Deaf people welcome the birth of deaf children; they find the concept of prenatal testing for a deaf child deeply disturbing; they oppose the use of cochlear implants and other technologies to overcome this sensory loss or difference.' A minority of deaf parents might one day go so far as to use genetic tests to make sure their children are born without hearing.

And how do you define 'undesirable' anyway? Disability is something that we all assume we know about, but which is hard to actually define in practice. 'There is fuzziness and dispute at the edges of the concept of disability,' says Jackie Scully, an ethicist at Newcastle University. 'At what point, for example, does an illness become disabling, and how should we distinguish people who are disabled through chronic illness from those who have an impairment but are not ill? What about people who are phenotypically anomalous but reject the suggestion that they are disabled? Too easy recourse to the umbrella term of disability without really being clear what it covers also glosses over the fact that what we find undesirable about disability – its disadvantage, distress or pain – can arise for a number of distinct reasons, not all of them biological.'

Should we worry about dysgenics?

Undesirable genes that lead to deviations from what we might call 'normal' may well spread. But thanks to genetic engineering and screening, so will the genes for desirable traits such as intellectual capacity, physical health and longevity, says Nick Bostrom. 'In any case, the time-scale for human natural genetic evolution seems much too grand for such developments to have any significant effect before other developments will have made the issue moot.'

ORGANIC CELL DISINTEGRATION

In the centre of each of our cells, stuck to the ends of the 46 strands of DNA that contain the instructions for building what we are, is a ticking clock. Every time a cell divides, this clock moves forward, each stroke bringing the cell closer to its eventual demise.

This clock is the telomere, a strand of DNA that caps the end of the chromosomes. Each division shortens the length of a cell's telomeres – very short telomeres lead to the death or disease of a cell, and this can cause a range of age-related diseases such as cancer, Alzheimer's, heart attacks and strokes. Telomeres are crucial failsafes for cells, preventing them from growing out of control and becoming cancerous.

But Reinhard Stindl of the Institute of Medical Biology at the University of Vienna thinks that this natural erosion of telomeres might not be restricted to individual cells. He thinks it might also be occurring in every successive generation of humans, and that this could lead to a point where the human population will crash. 'Over thousands of generations the telomere gets eroded down to its critical level. Once at the critical level we would expect to see outbreaks of age-related diseases occurring earlier in life and finally a population crash. Telomere erosion could explain the disappearance of a seemingly successful species, such as Neanderthal man, with no need for external factors such as climate change.'

Not just humans, either. Stindl believes that across all species, apart from bacteria and algae, telomeres are getting shorter with every generation. That means that an inbuilt evolutionary clock is ticking down to an inexorable extinction date for all complex life.

What is a telomere?

Deep in the nuclei of our cells lie twenty-three pairs of gossamer-like strands of DNA. The chromosomes, one of each pair coming from mum and the other from dad, contain genes that code for proteins spaced out by vast stretches of DNA that used to be called 'junk' but are now referred to simply as 'non-coding'. At the end of each chromosome is the telomere, a characteristic sequence of base-pairs present in everything from the simplest amoeba to humans, which acts as a full stop to the long strand of DNA.

Telomeres are there to protect the chromosomes, which might otherwise fuse in the nucleus or swap genetic material by accident. These sorts of mistakes can lead to cancer or other abnormal behaviour by the cell.

The end caps are also crucial in ensuring that cells can divide properly, and therefore in keeping a life form healthy. Every time a cell divides, its machinery has to copy all the base-pairs of DNA exactly, so that each daughter cell contains the full complement of genetic material. Unfortunately, this copying machinery can never get down to the ends of the DNA strands. If the telomeres were not there, the copying mechanism would snip off the ends of chromosomes every time they did their work. This would lead to the loss of crucial DNA (and its genetic information) every time a cell divided. Instead of snipping off crucial genes, then, the cell division process snips off a section of telomere.

The length of a telomere is a sign of the age of a cell. As cells divide or carry out their functions, they pick up mutations and errors in their DNA. These errors are all potentially dangerous; any of them could lead to cancer. When telomeres are short, it means that a cell is old and, at this stage, probably contains lots of DNA errors that have built up over time. Further divisions are prevented, as they would be potentially risky for the health of the organism as a whole. The cell enters a phase known as senescence and can divide no more.

Telomeres through the generations

Mass extinctions on Earth are well documented and, if not fully explained, there are plenty of ideas about how or why they might have occurred, including major environmental changes or asteroid impacts. For more than a century, though, scientists have been puzzled by the slower background rate of extinction of many of the species that have disappeared through the history of life on Earth. Well over 99 per cent of species that have ever existed are now no longer around, but only around 4 per cent of the loss can be explained by mass extinctions. Most species are not killed by asteroids or climate change; they just seem to whimper out.

The fossil record also shows that species go through fits and starts of change, rather than the gradual and continuous pace of evolution that might be expected from Charles Darwin's theory of natural selection. There are long periods in history where evolution just seems to stagnate.

Reinhard Stindl thinks he has the answer to both these conundrums. He believes that species have remained happily stable throughout the history of the Earth until a point where they have suffered a surprise population crash, caused by drastically short telomeres.

He discussed his idea in a 2004 paper for the *Journal of Experimental Zoology Part B: Molecular and Developmental Evolution*. 'The species clock hypothesis,' he wrote, 'is based on the idea of a tiny loss of mean telomere length per generation. This mechanism would not rapidly endanger the survival of a particular species. Yet, after many thousands of generations, critically short telomeres could lead to the weakening and even the extinction of old species and would simultaneously create the unstable chromosomal environment that might result in the origination of new species.'

This could be the case. The length of telomeres varies widely across species – in some birds it is up to a million base-pairs long, while in humans it is just over 10,000 base-pairs – and this might be indicative of Stindl's 'species clock'.

In the hypothesis, the erosion of the telomeres takes hundreds of thousands of years, which might explain the long, seemingly stagnant periods in evolution seen in the fossil record.

Is there a way to stop the decline?

If Stindl's ideas are correct, how can we deal with the ticking clock in our DNA?

The body naturally produces an enzyme called telomerase reverse transcriptase, which can rebuild telomeres at the end of chromosomes after cells have divided. It might be possible to lengthen telomeres in embryos by somehow increasing the activity of this enzyme, thereby giving that person's cells a longer lifespan. Unfortunately, adult humans produce little or none of it.

Scientists at the Dana-Farber Cancer Institute at Harvard Medical School have managed to rejuvenate mice by injecting them with telomerase. Without the enzyme the mice, which had been bred to mimic adult humans and not produce telomerase naturally, had been suffering from the effects of premature ageing, including a poor sense of smell, shrinking brains, infertility and damaged intestines and spleens. When the mice were injected with telomerase for a month, the signs of ageing were reversed. However, it is a proof of principle, and there would be problems translating that directly into humans – increased use of telomerase could lead to unchecked cell division, which is the start of cancer.

Telomeres can also lengthen in the normal course of things. After one of Stindl's population crashes, small pockets of animals would have carried on, and Stindl reckons that inbreeding among the groups of survivors could reset the biological clock and lengthen the telomeres once again. There is some evidence from laboratory mice to support this idea of resetting – they have been intensively bred from a small starting population and have very much longer telomeres than their wild cousins.

Another solution might be to meditate. In the Shamatha Project, scientists from the University of California, San Francisco, have been investigating whether telomeres can be affected by psychological factors. Their early results are intriguing: at the end of a three-month retreat of meditation and relaxation, attendants of the Shambhala Mountain Centre in Colorado had significantly raised telomerase activity. This is the first step to a reverse in cell ageing and, eventually, a lengthening of the body's telomeres.

It is unlikely that we will find a safe and robust way to lengthen our telomeres any time soon. But as the participants in the Shamatha Project might say, knowing the nature of the problem is the surest path to finding an answer.

THE FUTURE

IT'S ALL A DREAM

You wake with a start, your sight blurred and your head feeling groggy. As the fog clears, you begin to remember. That life you led, the one on Earth – the Earth you have known, with its billions of people, its scientific achievements, art, culture, politics, problems, horrors and beauty – was nothing more than a dream. You imagined the whole thing. You have no idea what is real.

Far-fetched or not, the idea that we are all in a dream has inspired philosophers and artists for thousands of years. It is a rich seam, from the ancient Chinese thinker Chuang Tsu, who wondered if he might be a butterfly dreaming he was a man, to multi-million-dollar Hollywood blockbusters such as *The Matrix*, in which a futuristic machine system imprisons all humans in a dream-like simulation of the modern world.

It is understandable. Dreams are the only immersive reality we experience apart from real life. When we dream, we do not know we are dreaming. However briefly, you really can believe that you are a Roman charioteer, a superhero capable of flying, or that you are having dinner with Albert Einstein. No one knows why we dream, but everyone has experienced a dream so vivid, they could not tell it was not real. It is, at the same time, the most natural, familiar and yet mystifying thing that billions of us do every day.

Artificial dreaming

If dreams happen naturally, is there any reason why we could not induce them artificially? Nick Bostrom, a philosopher and director of the Future of Humanity Institute at the University of Oxford, thinks it is just a matter of time. The basic idea behind his so-called 'simulation argument' is that the vast amounts of computing power that may become available through technological advances in the future could be used, among other things, to run large numbers of fine-grained simulations of past human civilizations.

This raises the startling possibility that our minds are, right now, the products of computer simulations at some point far in the future. A bit like the people plugged into the Matrix, we think we are living in 2011, but in actual fact it is 2311 in the real world and our minds are being fed a simulated past. 'If we are, we suffer the risk that the simulation may be shut down at any time,' says Bostrom. 'Until a refutation appears of the argument . . . it would be intellectually dishonest to neglect to mention simulation-shutdown as a potential extinction mode.'

Types of simulation

If you touch something hot, heat sensors in your fingers fire signals to your brain, which issues instructions telling you to pull your hand back via electrical signals to the muscles in your arm. The millions of inputs and outputs that characterize everyday actions are relegated to the subconscious, largely hidden from us as we go about our daily lives. Our thinking mind, the bit we are conscious of controlling, is aware of just a fraction of the inputs coming in from the senses, enough to keep us going without overloading us.

Our awareness of the world and our place in it is a result of electrical impulses that circulate around our brain cells. We interpret some of these electrical circuits as memories, others as feelings of pressure, yet more as elements of vision or sound. We exist in the world as flesh, blood and bone. But we only know it because of the electricity flowing around these brain circuits.

What if you could artificially induce these flows of electricity? Using implants, perhaps, what if a computer program could electrically stimulate your brain cells in precisely the right way at the right time to make your brain think your body was actually in a field surrounded by flowers, rather than strapped to a computer?

This is called the 'brain in a vat' scenario. The philosopher David J. Chalmers describes it thus: a 'disembodied brain is floating in a vat, inside a scientist's laboratory. The scientist has arranged that the brain will be stimulated with the same sort of inputs that a normal embodied brain receives. To do this, the brain is connected to a giant computer simulation of a world. The simulation determines which inputs the brain receives. When the brain produces outputs, these are fed back into the simulation. The internal state of the brain is just like that of a normal brain, despite the fact that it lacks a body. From the brain's point of view, things seem very much as they seem to you and me.'

The brain is clearly deluded by the simulation. It thinks it has a body, when none actually exists. It believes it is walking outside in a sunny park in London, when it is in fact in a dark lab in San Francisco. The reality experienced by the brain is clearly false, at least compared to 'real' reality. Nonetheless, the brain (and the person in that brain) experiences its world with as much visceral clarity as you or I do ours.

How do we know what is real?

'When the possibility of a matrix is raised, a question immediately follows. How do I know that I am not in a matrix? After all, there could be a brain in a vat structured exactly like my brain, hooked up to a matrix, with experiences indistinguishable from those I am having now,' says Chalmers. 'From the inside, there is no way to tell for sure that I am not in the situation of the brain in a vat. So it seems that there is no way to know for sure that I am not in a matrix.'

The point of a simulation is that you cannot tell that you are in it. If the designer of the simulation was benevolent and wanted us to know, however, he or she could make it abundantly clear what was happening. 'For example, the simulators could make a "window" pop up in front of you with the text "YOU ARE LIVING IN A COMPUTER SIMULATION. CLICK HERE FOR MORE INFORMATION." Or they could uplift you into their level of reality,' says Bostrom. Or we could find strong indirect evidence one day, perhaps by creating our own perfect simulation.

But what if our computer simulators were not feeling so benevolent? In that case, we had better hope that their simulation is not so perfect.

Perhaps the computing power needed to create a perfect simulated universe that behaves like ours at the smallest scales is too great, even for futuristic civilizations. But a simulation does not have to be anywhere near comprehensive in order to fool our brains. Your simulated hands do not need to have the structure of real hands, unless you are going to train a microscope on them, in which case the simulation could easily be added to for that particular moment. In other words, details only emerge if someone pays attention.

'The idea is that some of our experimental findings could be "faked" by the simulators, if they wanted to conceal the fact that we are in a simulation and if such faking was the most efficient way to conceal it,' says Bostrom. 'Consider the fact that while you are dreaming, your own brain often succeeds in making you unaware of the fact that you are dreaming. If your own humble brain can do this, then presumably it would be quite feasible for some technologically super-advanced builders of ancestor-simulations to achieve the same delusion.'

A more promising method of working out whether we are in a simulation is to look for glitches and bugs in the program. These might manifest themselves as miracles or paranormal activity in the world around us. Perhaps as things that defy the known laws of physics. Again, this strategy is fraught with problems: seeing something odd is not conclusive, and a motivated simulator could simply erase the mind of anyone in the simulation who was hell-bent on looking for errors.

Does it matter if we are dreaming?

We might be brains floating in jars somewhere in a dark room, but our beliefs are still always real, says Chalmers. Perhaps, he adds, we are disembodied minds outside space–time, made of ectoplasm. 'When I think "I am outside in the sun", an angel might look at my ectoplasmic mind and note that in fact it is not exposed to any sun at all. Does it follow that my thought is incorrect? Presumably not: I can be outside in the sun, even if my ectoplasmic mind is not. The angel would be wrong to infer that I have an incorrect belief. Likewise, we should not infer that an envatted being has an incorrect belief.'

The moral, he says, is that the immediate surroundings of our minds may well be irrelevant to the truth of most of our beliefs. What matters is the processes our minds are connected to, those sensory and motor inputs that create the illusion of reality in our brains. Everything else is extraneous anyway.

INFORMATION EXTINCTION

Many tens of thousands of years ago, groups of early humans would make marks on the walls and ceilings of caves. They scratched and painted images of bison and elephants and the people who hunted them. Some of them left behind abstract symbols on the rock, others made clear handprints. Thousands of years on, we can still see the marks they made.

Throughout time, humans have used many ways to record their knowledge and tell the stories of their lives. Only a few thousand years ago, Babylonians made records of the movements of heavenly bodies on clay tablets. And all of that information is intact. It may take time for us to understand what they are telling us, but we can see the symbols and work out something of the lives of our ancestors.

Thousands and tens of thousands of years from now, what will our descendants know about us? We have more knowledge today than any of our ancient ancestors, and we probably create more knowledge in a year than our predecessors might have done in centuries. We record all of it in myriad ways – on paper, in painting, but mostly on electronic hard disks.

What if the Earth were to succumb to some cataclysm in which the computers that store our knowledge were all compromised? Perhaps the power grids would go down for good, or most of the world's population would be wiped out by a virus, leaving no one left to maintain the machines. If, in tens of thousands of years, we wanted to tell the story of the 21st century, what would our descendants be able to retrieve?

If they forgot all about us, it might not be the end of the world in physical terms, but it would be the end of *our* world.

Our modern memories

We live in an unparalleled age in human history, driven by information and kept functioning through an intricate network of computers, switches and data connections. It is the latest step in the human ability to transfer complex knowledge from one person to another, from one place to another, from one generation to another, an ability that has defined our path to becoming the most dominant species on the planet.

To get to where we are, we needed a collective memory, something that all of humanity could add to and access. First there were stone tablets distributed among the few, then came paper and writing, after that books, and today, we have hard disks. Whereas in the past the systematic records in existence were bureaucratic, legal or ecclesiastical documents, nowadays we record much more and more often.

'The digital revolution is transforming the nature of personal archiving: from curation techniques to the kinds of lives being preserved for posterity – not just the rich and famous, but now everyone participating in the digital age,' says Jeremy Leighton John, curator of eMANUSCRIPTS at the British Library. With the emergence of personal computing in the

1970s, says John, more and more people are passing on details of their lives to future generations as digital files. The electronic hard disk is the substrate for our collective memory, and there are several in each modern home, hundreds in each scientific laboratory, billions around the world.

On those disks are all of our ideas – histories, photographs, diaries, data, novels, bank records, films, scientific theorems – whole or in fragments. And most of them are personal: the International Data Corporation estimated that by the end of the first decade of the 21st century, around 70 per cent of the world's digital information will be created by individuals rather than organizations.

The problem is that hard disks have never been intended for very long-term storage and no one really knows if they will survive with their precious cargo intact. Every year, as technology advances, hard disks become smaller, thinner and more dense with information. This is good if you want a smaller computer or to put memory in places where it has never been before (a fridge, say, or a phone). But there are problems, too – the smaller a disk is, the more information is kept on it per square centimetre and the more of that information you lose if the disk gets damaged.

Important information on hard disks is often backed up on magnetic tapes or optical discs such as CDs or DVDs. But these formats are no more trustworthy than hard disks. In tests, the best discs have been shown to survive for perhaps a century. The cheapest CDs will last between five and ten years before dropping information and starting to become unreliable.

And the tide of new information goes on. 'Future generations may well have comprehensive video and [GPS] recordings

of their lives, along with records of neurological and physiological parameters such as heart rate, not to mention personal
DNA sequences,' says John. 'Bionic devices that partially restore,
enhance or extend an individual's physical or sensory capabilities will be digitally tuned for each individual, just as digital
hearing aids are today. Personal fabricators will allow individuals to create for themselves useful and ornamental physical
artefacts. People already interact online through virtual versions of themselves in online gaming environments, and in
future this will be advanced with immersive visualization,
complemented by touch, taste and scent. What will happen to
these digital representations in the long run?'

How much of a loss would it be if this information was
irretrievable in the future? The details of your daily life via
Twitter, Facebook or Flickr probably will not matter much to
the humans living 20,000 years from now. It is the same with
the surviving ancient manuscripts and stone tablets – these are
important for their historic rather than their informational value.

But what about all that genome-sequencing data that we
have been patiently collecting for the past few decades? How
about the raw information from the Large Hadron Collider,
which can be mined for years to come to understand and
explain the laws of physics? How about the blueprints to make
aircraft, computer circuits and radio transmitters? What about
our great novels and the histories of what happened while we
were alive?

Sending messages to the future

Another problem presents itself to anyone thinking about how
to make sure knowledge persists: language. Would we be able

to understand the humans living 10,000 years ago? Probably not. So who is to say that anyone in the future will speak our language? It is a problem facing scientists who want to bury the waste from nuclear power stations, material that will remain radioactive for hundreds of thousands of years. They need to place signs outside the facilities that will communicate the danger of what lies within, and there is no reason to expect that humans 500 generations from now will understand our language or customs. The nuclear scientists have to guess what might be meaningful to people in the distant future.

Even today, talking across different countries can be difficult, never mind thinking about thousands of years in the future. Red means danger in one culture but luck in another; a creepy-crawlie might be dirty or scary in one place but be seen as a tasty snack in another. The scientists' solution to the nuclear conundrum is to use facial expressions: an image adapted from the famous painting *The Scream* by Edvard Munch is thought to mean the same thing to everyone now and (hopefully) in the future too.

Scientific legacy

Maintaining archives of data for the future has only recently become a priority for scientific institutions. And not just for curiosity's sake: in future, scientists will need access to raw data so that they can perform brand-new analyses or even look for proof of new theorems or evidence of scientific fraud in old data.

Take the LHC in Geneva. It will generate almost 500 million gigabytes of data in its fifteen-year run of experiments, and that data will be stored on disks and tapes and distributed to the world's scientists via a 'grid' of 100,000 computers. The information technology project to allow the collection and management of all this data at Cern has been decades in the planning, but there is little idea of how the information will be kept in storage beyond the lifetime of the LHC itself, which will shut down early in the 2020s.

One way to keep data more resilient for longer is to copy it from one disk or tape to another. However, doing that systematically for immense data sets (such as the one from the LHC or from gene-sequencing projects) increases the risk of errors creeping into the information.

Individuals also copy files to and from personal computers. It may well be the case that in the future, there is more chance of our descendants coming across a Michael Jackson album on a historical hard disk than the latest papers on general relativity or details about how to extract metals from the Earth or make important drugs.

The other failsafe is an old one: store more information on paper. The oldest surviving book, found in a cave in China, dates back to the 9th century AD. If books are kept in stable conditions and away from hungry pests, there is no reason why they (and their information) cannot survive for a thousand years or more.

A more systematic approach has been thought up by the the Long Now Foundation, a California-based organization. Its alternative to books is the 'Rosetta Disk', which is made from nickel and holds descriptions of 1,000 languages. On one side it is etched with text that starts off at readable size and then goes down to the nanoscale (billionths of a metre across). On the reverse of the disk are up to 14,000 pages of text, viewable with an enclosed magnifying glass. The foundation reckon that the disk could remain legible for thousands of years.

We revel in information today. The ease with which our lives are connected to each other and to the vast resources of the World Wide Web has given us a sense of security that perhaps it will always be like this. What we keep in our memories has changed in this electronic world, and not necessarily for the worse – we have access to much more knowledge today than our parents and grandparents ever had. But as we change how we remember, we need to be aware of (and prepare for) what it would be like to suddenly forget.

UNKNOWN UNKNOWNS

We don't know what we don't know. The geniuses of the past would never have guessed at nuclear war or climate change. And the geniuses of today cannot know what knowledge (and the allied risks) will emerge in the future.

Five months after the terrorist attack on September 11, 2001, had claimed thousands of lives at the World Trade Center in New York and the Pentagon in Virginia, US secretary of defense Donald Rumsfeld was speaking to members of the press about the various threats facing America. He had been talking specifically about the regime in Iraq when a reporter put up their hand and the following interchange took place:

Q: *In regard to Iraq, weapons of mass destruction and terrorists, is there any evidence to indicate that Iraq has attempted to or is willing to supply terrorists with weapons of mass destruction? Because there are reports that there is no evidence of a direct link between Baghdad and some of these terrorist organizations.*

Rumsfeld: *Reports that say that something hasn't happened are always interesting to me, because as we know, there are known knowns; there are things we know we know. We also know there are known unknowns; that is to say we know there are some things we do not know. But there are also unknown unknowns – the ones we don't know we don't know. And if one looks throughout the history of our country and other free countries, it is the latter category that tend to be the difficult ones.*

Rumsfeld was given a lot of stick for his 'unknown unknowns' remark. A tautology, perhaps? Management-speak bent to the worst degree? Not quite. When assessing danger and catastrophe, no conversation can be truly complete without thinking the unknown. And within that unknown, it would be the height of arrogance and hubris not to admit that there are things you know you don't know and, inevitably, any number of things that you don't know you don't know.

'We need a catch-all category,' says philosopher Nick Bostrom. 'It would be foolish to be confident that we have already imagined and anticipated all significant risks. Future technological or scientific developments may very well reveal novel ways of destroying the world.'

Predicting the unpredictable

Take the most learned people during the Enlightenment period in western Europe – Isaac Newton, say, Francis Bacon or Bishop George Berkeley – and imagine asking them how they thought the world might come to an end. There would be tales of divine intervention (Newton believed that doomsday would happen in the 21st century, calculated from clues in the Bible), and someone might offer the idea of a bloody war causing so many casualties that nations would suffer and wither away. There might be serious consideration of other fantastical theories too, but none of these clever people could have told you about the doomsday potential of a nuclear bomb. Or an asteroid strike. Or rising sea levels due to climate change.

Three hundred years ago, we knew less science and much less about the world and universe around us. With knowledge comes power and, inevitably, the understanding that all around us there is new danger.

With cosmic threats to our existence, the danger has been there all along; it just took us some time to notice it: the collision of our galaxy with Andromeda, for example, or the arrival of a black hole. Common to all cosmic threats is that there is very little we can do about them even when we know the danger exists, except to try and work out how to survive the aftermath.

Other threats are a matter of self-control. These are dangers we have made for ourselves by applying our brains, working out how to manipulate the world and using that knowledge for our own ends – nuclear weapons and climate change are good examples. In both cases, any risks have been caused by, and can also be prevented by, the action of people.

Applying our brains is also where new, unforeseen problems for the world could turn up. How do we start to think about this, though?

One group used to working out and predicting the unthinkable is the military. At the Pentagon, commanders talk about 'disruptive threats' – weapons or tactics that come from nowhere to tilt the balance of power during combat. These have included the earlier-than-expected detonation of the first Soviet nuclear bomb (in 1949) and the launch of Sputnik 1, the world's first orbiting satellite. Pentagon and White House analysts have identified future disruptive threats as, for example, weapons that could use 'biotechnology, cyber and space operations, or directed energy weapons' or ones that could reliably shoot down missiles and warplanes from an invading force.

In the information age, predicting the next source of big problems becomes even more complex. The US might have worried about nuclear armageddon in the past, but it always knew that it would take the Soviets lots of manpower and years of testing to get a sizeable arsenal of warheads together. Today's wars can happen online and they can happen fast; there's no waiting around to build and test expensive new weapons. In the online doomsday scenario, the military might not find out they are being attacked until it is too late to do anything about it.

Black Swans

There is another way to think about unknown unknowns. In his book *The Black Swan*, the economist Nassim Nicholas Taleb wrote about the idea of big surprises that end up having major impacts. He highlighted the importance, through history, of events that were hard to predict, had a huge impact, were rare and which went beyond the normal expectations of history, science or economics. Included in this list of so-called Black Swan events are the rise of the internet, the First World War and the 9/11 terrorist attacks. Taleb also includes almost every scientific discovery and major artistic achievement.

'Before the discovery of Australia, people in the old world were convinced that *all* swans were white, an unassailable belief as it seemed completely confirmed by empirical evidence,' he wrote. 'The sighting of the first black swan might have been an interesting surprise for a few ornithologists (and others extremely concerned with the coloring of birds), but that is not where the significance of the story lies. It illustrates a severe limitation to our learning from observations or experience and the fragility of our knowledge. One single observation can invalidate a general statement derived from millennia of confirmatory sightings of millions of white swans. All you need is one single (and, I am told, quite ugly) black bird.'

He generalizes the Black Swan event as something with three attributes. First, it is an outlier, something beyond our normal expectations and for which the past is not a reliable guide. Second, it has an extreme impact. And third, it is something that makes human nature rationalize the event afterwards.

'A small number of Black Swans explain almost everything in our world, from the success of ideas and religions, to the dynamics of historical events, to elements of our own personal lives,' says Taleb. 'Ever since we left the Pleistocene, some ten millennia ago, the effect of these Black Swans has been increasing. It started accelerating during the industrial revolution, as the world started getting more complicated, while ordinary events, the ones we study and discuss and try to predict from reading the newspapers, have become increasingly inconsequential.'

Think of 9/11, he says. Had the risk of an atrocity been reasonably conceivable the day before, the whole thing would not have happened. 'If such a possibility were deemed worthy of attention, fighter planes would have circled the sky above the twin towers, airplanes would have had locked bulletproof doors, and the attack would not have taken place, period. Something else might have taken place. What? I don't know. Isn't it strange to see an event happening precisely because it was not supposed to happen? What kind of defense do we have against that? Whatever you come to know (that New York is an easy terrorist target, for instance) may become inconsequential if your enemy knows that you know it. It may be odd to realize that, in such a strategic game, what you know can be truly inconsequential.'

This also works in businesses and scientific ideas – any 'secret recipe' for setting up a successful restaurant would spread like wildfire and everyone in the street would be using it; the real big idea would be something that most restaurateurs had not yet conceived of. 'The same applies to the shoe and the book businesses – or any kind of entrepreneurship. The same applies to scientific theories – nobody has interest in listening to trivialities. The payoff of a human venture is, in general, inversely proportional to what it is expected to be.'

If the Pacific tsunami of December 2004 had been expected, it would not have caused the damage it did, says Taleb. 'The areas affected would have been less populated, an early warning system would have been put in place. What you know cannot really hurt you.'

Despite that, we all behave as though we are somehow able to predict events based on our knowledge and history. 'We produce thirty year projections of social security deficits and oil prices without realizing that we cannot even predict these for next summer – our cumulative prediction errors for political and economic events are so monstrous that every time I look at the empirical record I have to pinch myself to verify that I am not dreaming. What is surprising is not the magnitude of our forecast errors, but our absence of awareness of it.'

One answer is to become aware of our inability to predict everything and be ready for *something* unpredictable to happen. The left-field Black Swan events of the past are not only shocking and impactful, they tell us a lot about human behaviour, about how we learn from history. When asking the question about unforeseen things that might bring the world to its knees, perhaps we need to step back and think about the question in a more reflexive way.

'We do not spontaneously learn that *we don't learn that we don't learn*. The problem lies in the structure of our minds: we don't learn rules, just facts, and only facts. Metarules (such as the rule that we have a tendency to not learn rules) we don't seem to be good at getting. We scorn the abstract; we scorn it with passion,' says Taleb.

This behaviour lies deep in our animal past – being thoughtful and introspective is no use to an animal on the savannah of early Africa if all he needs to do is notice and run away from lions. 'Consider that thinking is time-consuming and generally a great waste of energy, that our predecessors spent more than a hundred million years as nonthinking mammals and that in the blip in our history during which we have used our brain we have used it on subjects too peripheral to matter. Evidence shows that we do much less thinking than we believe we do – except, of course, when we think about it.'

There is no answer to expecting the unexpected, no discrete way. But taking a lesson from Taleb, it is better to know that and use the information. In thinking about the unexpected ways in which the world could end, therefore, the past is no guide at all.

INDEX

INDEX

INDEX

INDEX